Digital Quality Management in Construction

Much has been written about Building Information Modelling (BIM) driving collaboration and innovation, but how will future quality managers and engineers develop digital capabilities in augmented and video realities, with business intelligence platforms, robots, new materials, artificial intelligence, blockchains, drones, laser scanning, data trusts, 3D printing and many other types of technological advances in construction? These emerging technologies are potential game changers that require new skills and processes.

Digital Quality Management in Construction is the first 'how to' book on harnessing novel disruptive technology in construction quality management. The book takes a tour of the new technologies and relates them to the management of quality, but also sets out a road map to build on proven lean construction techniques and embed technologically based processes to raise quality professionals' digital capabilities. With the mountain of data being generated, quality managers need to unlock its value to drive the quality of construction in the twenty-first century, and this book will help them do that and allow those working in construction quality management to survive and thrive, creating higher quality levels and less waste.

This book is essential reading for quality managers, project managers and all professionals in the Architecture, Engineering and Construction industry (AEC). Students interested in new and disruptive technologies will also learn a great deal from reading this book, written by a professional quality manager with nearly thirty years' experience in both the public and private sectors.

Paul Marsden is a Chartered Quality Professional with the Chartered Quality Institute, who has nearly thirty years' experience of quality management in the construction, telecommunications, banking, security, aerospace, energy and rail fields. He specialises in quality management capability, sustainability, continual improvement, customer satisfaction, lean programmes, auditing and creating business intelligence platforms to collate and report performance data and information. Paul has had a diverse career, including being a Member of Parliament, Head of Quality at Horizon Nuclear Power, Head of Quality at Horizon Nuclear Power and Interim Secretary-General of a European construction trade association.

Digital Quality Management in Construction

Paul Marsden

LONDON AND NEW YORK

First published 2019
by Routledge
2 Park Square, Milton Park, Abingdon, Oxon OX14 4RN

and by Routledge
52 Vanderbilt Avenue, New York, NY 10017

Routledge is an imprint of the Taylor & Francis Group, an informa business

British Library Cataloguing-in-Publication Data
A catalogue record for this book is available from the British Library

Library of Congress Cataloging-in-Publication Data
Names: Marsden, Paul, 1968 – author.
Title: Digital quality management in construction/Paul Marsden.
Description: Abingdon, Oxon; New York, NY: Routledge is an imprint of
the Taylor & Francis Group, an Informa Business, 2019. | Includes
bibliographical references.
Identifiers: LCCN 2018056412 | ISBN 9781138390799 (hbk) |
ISBN 9781138390829 (pbk) | ISBN 9780429423062 (ebk)
Subjects: LCSH: Building–Quality control–Data processing. |
Building–Quality control–History.
Classification: LCC TH438.2.M37 2019 | DDC 624.0285–dc23
LC record available at https://lccn.loc.gov/2018056412

ISBN: 978-1-138-39079-9 (hbk)
ISBN: 978-1-138-39082-9 (pbk)
ISBN: 978-0-429-42306-2 (ebk)

Typeset in Goudy
by Wearset Ltd, Boldon, Tyne and Wear
Printed by CPI Group (UK) Ltd, Croydon CR0 4YY

Contents

Figures

viii *Figures*

Tables

Preface

When I look back at my own career in construction, I can see that some of the quality control tools I was trained to use were essentially the same as those in use back in the 1960s. Inspection and Test Plans and the quality management systems are still based on the quality standards introduced in the 1980s, such as ISO 9001, with Project Quality Plans and internal audits, albeit with a new emphasis on risk management.

Choosing civil engineering as a degree course only came about because, when I was fifteen years old, I sat through an entertaining presentation in the school hall by a local college lecturer on why 'A' Levels were not for everyone and there was an alternative. The thought of even more concentrated doses of physics or maths while I was at that moment ploughing through concentrated doses of physics and maths at 'O' Level was almost enough to make me give up on all studying, so any kind of viable alternative made me prick my ears up.

I mostly remember the invited lecturer's whimsical stories about getting out of being given a speeding ticket by jumping out of his car and using psychology as he rested his arm on the police officer's car and leant in through the open window to beg for forgiveness. The startled police officer, so used to getting out of the car and strolling down the road to a trembling driver, actually accepted the apology and let the lecturer off without a ticket. With such engaging but totally irrelevant stories by the lecturer in my head, I somehow was persuaded to sign up for a Diploma in Building course at Mid-Cheshire College of Further Education. And, yes, the connection of speeding tickets and building studies still eludes me to this day.

I thoroughly enjoyed studying for the diploma since there were practical course units on bricklaying, electrics and woodwork. I also took an additional 'O' Level in computer studies at college and the tedium of computer programming was offset by wiring plugs and plastering. I bounded through the building diploma with distinctions and naturally after learning all about house building, the logical sequel was a degree in civil engineering at Teesside Polytechnic. Moving from houses to roads, dams and bridges, I thought would involve more of the same hands-on learning. Little did I realise that the degree would involve sitting through stiflingly long lectures on hydraulics, unfathomable mathematical formulae and endless hours sitting in the library figuring out all the stuff not taught to us in the lectures. Politics distracted me at polytechnic (which again interrupted my career in construction in the late 1990s), and I never did finish my degree.

However, returning to the point of this story, when I was at Further Education college, learning practical, real-world things (that were actually useful in work and at home), I was shown some basic inspection and testing for concrete, such as slump tests and cube tests. How to test the strength of concrete all seemed straightforward: fill a metal box, tamp down the layers of liquid concrete, wait a day before removing from the mould and then carefully store before being tested by a calibrated compression machine to work out its strength. Likewise, taking a tape measure to check on the size and accuracy of a floor slab seemed a touch too late to do much about any errors found, once the concrete or flooring had been laid.

It was not until I was making the cubes 'in anger' as a young, unqualified civil engineer working for Taylor Woodrow Construction at the Wanlip Water Treatment Works construction site, near Leicester, in 1990, on my first construction project after polytechnic that it occurred to me that I was taking samples from concrete that was also being poured in front of me into formwork to create large reinforced walls as part of the tank. The cubes would not be tested for seven days or more *after* the huge quantities had filled the formwork and started to cure. It seemed rather late in the day to be testing.

Sure enough, I unfortunately proved the point myself, when I was left in charge of setting out a long reinforced concrete wall, which was part of the water treatment works, one summer's evening after most of the site staff had gone. I was left with a rather irritable subcontractor, who also wanted to get away to watch Italia '90. The next day, it was perfectly apparent that the job was botched and we didn't need to wait for the cube results to see that a jack hammer would be needed to remove sections of reinforced concrete from my error-strewn setting-out.

After an urgent site meeting was called to look at the bent, poorly formed wall, I was quietly shuffled off the Wanlip project out of harm's way to Yorkshire on a surveying project. I always thought that it made better sense to test before concrete was poured or find a different way with technology, to get the construction results more quickly.

I had been surveying sewers in Yorkshire for six months when I started complaining to the contracts manager. He seemed a bit perplexed why I wasn't happy, although as a civil engineer fresh out of polytechnic who thought he would be building roads and bridges, it should not have been too much of a surprise that I was fed up breaking my back, lifting up manholes day after day and pouring coloured paint powder down them to see where the drains met the sewers. It was bad enough for me, but for my surveying partner, Alan, who had been a proper civil engineer for many years actually building things, it was a much greater insult.

When we were first introduced, Alan had said that we would take it in turns lifting the manhole covers, while the other one used a scale ruler and scratched out lines in pencil on a printed map of where we thought the drains ran from the houses to join the sewers in the roads. He suggested that I go first at the messy end of the job. Six months later, I was still asking when it would be my turn drawing the pictures.

Alan and I were one of just three teams roaming around the streets of Huddersfield in 1991, systematically finding what was termed Section 24 sewers that had not been previously recorded and, due to new regulations, had opened up the water company to potential liabilities. In the past, water inspectors had carried such knowledge around in their heads, i.e. which drains and sewers were the authority's responsibilities and which were the householders' responsibilities. Hence, Yorkshire Water had employed Taylor Woodrow on a pilot study to start to assess the scale of the task. It triggered a lot of smirking and merriment among our colleagues on other projects, who were typically calculating shearing stress in skyscrapers' structural steel beams or something similar to update their three-dimensional (3D) computer-aided design (CAD) drawings on the computer, when they heard we were using hammers and chisels to wrench up filthy manhole covers that had lain undisturbed for decades, gagging on the stench as we dropped tape measures down them to create A4 sheets of paper with coloured lines of the survey.

I had had enough of stomping up and down dale, carrying a bag of tools in the icy weather for ten hours a day, munching on cheese and onion pasties for lunch, huddled in a lime green van with the heater on full blast while Alan tried to interpret his own handwriting showing the measurements and notes on sizes of pipes. In those days, it was instant coffee from a flask or a cafe to thaw out our frozen fingers.

The contracts manager duly noted my complaints and promptly ignored them but the project manager was starting to get a bit of grief from the client, Roger at Yorkshire Water, about the quality of the work that was being produced.

One day, I was summoned to our office in Leeds and sat in the project manager's tiny office as he slid a copy of British Standard BS 5750, across the desk to me. I concentrated hard as I flicked through it, wondering what I was supposed to do with it.

With his feet on the desk, supping his tea from a chipped mug, he explained that the client was not satisfied with the quality of the maps.

'Have a look at that. We need a quality management system based on BS 5750 and that means we need someone to write a manual. That should keep the client happy.' I nodded, without understanding a word he had said.

'Does that mean I don't need to go back out on to site?' 'Site' being a euphemism for the streets of Huddersfield.

I must have had a huge smile on my face as it dawned on me that writing would entail sitting in a warm office.

'Yes, but it's only a couple of weeks' work.'

I must have looked crestfallen as I left his office clutching the standard and wondering what I had got myself into. That was my introduction to quality management and, between diverging into politics for a while, that has been my career for the past thirty years.

Essentially, the mechanics of the discipline have not changed since the 1960s. The standard set out the framework for getting key activities right that would help deliver products and services to conform with a specification to satisfy a customer. The quality manager was the one to create and manage the management system and audit conformance to it. The principles are sound, though too often in construction, the creation of policies and procedures are done to achieve a certificate based on a tick box approach, undermining the credibility of the standard and utterly failing to improve the performance of the organisation.

Our little contract in Yorkshire became a modest success over the next two years and led to a prototype tablet computer for storing Ordinance Survey maps and recording the lines of the sewers. That survey tablet in Yorkshire was my first exposure to digital information management. I was amazed that hundreds of A4 paper maps covering the whole of Yorkshire could be stored on one computer the size of a book.

I next encountered quality management on a construction site (a proper one this time) at Fleetwood in the north-west of England. developing a water treatment works to help clean up the Fylde coast beaches. Curiously, the locals did not appreciate sewage being pumped straight out to sea.

I was appointed the Project Quality Manager and had to design a quality management system with inspection and testing plans overviewing a long supply chain of material suppliers and subcontractors. It was generally deemed a necessary evil that the site teams had to fill in lots of forms (lots and lots) to record an accountable audit trail of evidence of the quality of work. But it also required the construction professionals to write down procedures, since few existed. I soon noticed that, given all the day-to-day pressures on quantity surveyors, site managers, foremen, planners, estimators, designers and a plethora of engineers, trying to get them to remember to write procedures and then follow their own procedures was a monumental task. Invariably I found that staff wanted to be helpful and understood the importance of the theory of quality management but the technical demands took precedence, until there was a crisis or problem. Then it was time to run around like idiots putting right the problems and promising to ensure there would be better quality assurance. Everyone would quietly forget about the promises and go back to their same old bad ways until the next crisis. Frankly, I don't think construction has changed much in its philosophy over the past thirty years.

I learnt how to master the so-called softer skills of building relationships, negotiating and communications. Holding a clipboard and adopting the inspector's air of superiority were the perfect combination to hack off construction people, and unless you enjoyed doors being slammed in your face and people shouting at you, then getting alongside the construction workers and listening to their problems and complaints usually was a better way to persuade them to engage. If they owned and used the quality management system, then there was a better chance that quality issues would be found and eradicated before they became problems.

I also realised the cumbersome bureaucratic nature of quality management created and maintained on paper systems. In those days a quality manual, procedures and forms would be created by the quality manager and then reprinted and individually numbered before being disseminated by post (or hand-delivered) to the great and the good. The 'great' were the directors who would give the manual to a secretary who would put it on a shelf and immediately forget about it. Those on site (the 'good') would be expected to sign a form to prove that they had read it, which would be posted back to the quality manager, who would inevitably have to send out reminders and make phone calls to chase up missing forms. Then, with each update to the management system, which could be a handful of grammatical errors or regulatory changes, the relevant pages would be reprinted and sent out to the designated document holder. Several times a year, I would waste a significant amount of time on producing, printing, photocopying and posting manuals and updates, and hence this was a bureaucratic task that negatively impacted on the reputation of construction quality management. I lost count of the number of times on an audit I found a project manager with a clutch of unopened envelopes from me stuffed into their pending tray.

The actual results of managing the quality, while inherently saving time, money and improving performance, always seemed academic to many and when I was challenged by others in the industry on why we needed quality assurance (QA), I resorted to time-honoured arguments that it was needed to keep the client happy and the procedures would ensure the right things were done at the right time, eliciting a bored nod and a wry smile. The organised chaos on many sites soon showed me the futility of an average quality manager in trying to convince directors to do the 'right thing'.

I recognise that, usually, written texts on the subject of construction are authored by men and pay tribute to men. However, there is no doubt that women have played equally important roles throughout history in designing, labouring, problem solving and leading in construction, but through misogyny and historical prejudices have not received the credit they deserve. I salute those 'forgotten' women whose ideas, knowledge and expertise have been embedded in the evolution of Quality Management, even though individuals may not have been recognised for their contribution. Trailblazers, whose names are known, include: Elmina Wilson (1870–1918), who was the first American woman to complete a civil engineering degree course and designed the Marston Water Tower, the first such tower built in steel, west of the Mississippi River. Later she worked on the Metropolitan Life Insurance Company Tower in New York. Likewise, Hattie Scott (1913–1993) was the first African-American woman to graduate as a civil engineer and later worked for the U.S. Army Corps of Engineers (USACE).

As the complexity of construction has increased with architectural and design demands, more regulations, new materials and client pressures to speed up the programme and reduce costs, so the level of information and data has likewise increased. Since construction quality management has been about collating and reporting on performance, so the proportion of such reporting versus the volume of project information being generated, has fallen to a tiny amount. As quality professionals, we barely scratch the surface of what is being created in data and information and cannot possibly be providing a fair representation of the actual performance. As such, quality management reporting can be misleading and providing a false comfort to project leaders.

Digital quality management recognises that we need to prioritise the data and information being analysed and report on the risks, to provide a better level of assurance to decision-makers. To achieve such analysis we need to use the technologies available to dig deeper and smarter into the data since the human mind alone cannot process the size of data and information that typically is created on construction projects.

It is time for a reappraisal not just of what quality management in construction needs to deliver but also the methods we use, given the level and pace of change, and that is the aim behind this book.

With printed versions of this book available in 2019, having been written in 2018, based on a full year of research up to 2017, then it is very difficult to provide an up-to-date commentary on all technological changes and their impacts. If some technologies or products have since been discontinued or significantly changed, then it simply shows the wider problem of publishing in a traditional paper format that cannot adjust to the pace of change. However, I hope that my approach to Digital Quality Management will help to generate a new discussion and debate within the profession and the wider construction industry on how to invest, research and develop new technologies that add value to construction by lowering risk and especially to make it a safe environment for humans to work in. I welcome readers' feedback and suggestions for improvements and I can be reached through Skype @paulwbmarsden or by email at paul.marsden1968@gmail.com.

Paul Marsden
October 2018

Acknowledgements

I know that this is the page that many readers will naturally skip over but please bear with me. Before you read and appreciate this book, then it is important to remember that there are many people who have made it possible, through their support, thoughtfulness, intellect and love and they deserve a minute of your time.

Those with whom I have worked since 1990 – and you know who you are – in many companies in construction, telecommunications, security, aerospace and defence industries; I thank you, as I have learnt so much from you and I hope I have added value in some humble way to those businesses.

Special appreciation goes to Alan Neal from my days at Taylor Woodrow and Colin Ellam at Horizon Nuclear Power, who provided me with huge learning opportunities during my career.

Recognition goes to the Chartered Quality Institute (CQI), who deliver important services to quality management and set the bar in global professional standards.

I would like to acknowledge the permission of Professor S. D. Lambert to reproduce material from Attic Inscriptions Online (www.atticinscriptions.com) for the translations in Chapter 2.

I also want to thank all those who gave their time to be interviewed by me and provide insights into best practices and innovative ways of working in quality management.

I thank my editor, Ed Needle and his team, especially Patrick Hetherington, for their support, advice and unstinting belief that this was a book worth publishing. My appreciation goes to Emma Critchley, Project Manager at Wearset and Susan Dunsmore, my copy-editor, who have patiently and professionally reviewed and finessed my manuscript.

I have spent many days at the Jaunty Goat café in Chester, enjoying their delicious coffee and their complementary Wi-fi, which has provided a welcome inspiration when writing days have been tough.

To my family – my mum, my sister, Pamela, my children, Alex, Richard and Luba, for putting up with absences from the dining table, bad moods and missed special occasions due to work; I apologise and thank you so much for your patience and fortitude. You have provided me with daily inspiration to keep going at the most difficult of times and given me so much joy and pride in your achievements. I also have learnt to appreciate that family is the most important part of life.

I thank most of all, my dear wife, Elena, for her intelligence, love, support and affection, without which this book would never have been written. Я люблю тебя всегда
Russian Cyrillic for 'I love you forever'.

Abbreviations

3D	three-dimensional
4D	four-dimensional
8D	eight disciplines problem solving model
AEC	Architecture, Engineering and Construction industry
AGI	Artificial General Intelligence
AHS	autonomous haulage system
AI	Artificial Intelligence
AIS	Advanced Industrial Science and Technology Institute
AlON	aluminium, oxygen, and nitrogen
AM.NUS	Additive Manufacturing at the National University of Singapore
ANI	Artificial Narrow Intelligence
ANN	artificial neural network
APPGEBE	All Party Parliamentary Group for Excellence in the Built Environment
AR	Augmented Reality
ASI	Artificial Super Intelligence
ASIM	Advanced Step in Innovative Mobility
ATL	autonomous track loader
BBA	British Board of Agrément
BEP	BIM Execution Plan
BI	Business intelligence
BIM	Building Information Modelling
BMS	business management system
BRE	British Research Establishment
BS	British Standard
BSI	British Standards Institution
BSRIA	Building Services Research and Information Association
CAA	Civil Aviation Authority
CAD	computer-assisted design
CAPEX	capital expenditure
CARS	credibility, accuracy, reasonableness and support
CAV	connected and autonomous vehicles
CDE	common data environment

CI	continual improvement
CIO	chief information officer
CIOB	Chartered Institute of Building
CoEs	Communities of Experience
col.	column
CONSig	Construction Special Interest Group
CoPs	Communities of Practice
CPP	Construction Phase Plan
CQI	Chartered Quality Institute
CSCS	Construction Skills Certification Scheme
CSR	corporate social responsibility
CSTB	Centre Sciéntifique et Technique du Bâtiment
DCMS	Digital Construction Management System
DfMA	Design for Manufacture and Assembly
DoE	Design of Experiments
EDI	Electronic Data Interchange
EDM	electronic distance measurement
EMP	Environmental Management Plan
FAA	Federal Aviation Administration
FMB	Federation of Master Builders
FMEA	Failure Modes and Effects Analysis
FOV	Field of View
FPV	First Person View
FTA	Fault Tree Analysis
GDP	Gross Domestic Product
GIS	geographic information system
GPS	Global Positioning System
H&S	Health and Safety
HUD	heads up display
IFC	Industry Foundation Classes
IMS	Integrated Management System
IoT	Internet of Things
IPCC	Intergovernmental Panel on Climate Change
IS	Information Systems
ISO	International Organization for Standardization
IT	Information Technology
ITP	Inspection and Test Plan
JCT	Joint Contract Tribunal
JISC	Joint Information Systems Committee
KM	knowledge management
KPI	key performance indicator
LIDAR	light imaging, detection and ranging
LPS	Last Planner System
LSS	Lean Six Sigma
M&E	mechanical and electrical

MDM	Master Data Management
MEWP	mobile elevating work platform
ML	Machine Learning
MR	Mixed Reality
MR	Management Review
MSE	Mean Square Error
NC	non-conformance
NCCR	National Centre of Competence in Research
NEC	New Engineering Contract
NGO	non-governmental organisation
NHBC	National House Building Council
NQEs	National Qualified Entities
OD	organisational development
OPEX	operating expenditure
OSC	Operating Safety Case
PDC	Process Development Committee
PfCO	permission for commercial operations
PO	process owner
PPE	personal protective equipment
PQP	project quality plan
PRA	probabilistic risk assessment
QA	quality assurance
QFD	quality function deployment
QMC	Quality Management Committee
QMS	quality management system
QSRMC	Quality Scheme for Ready Mixed Concrete
RACI	responsible-accountable-consulted-informed
R&D	research and development
RCA	root cause analysis
RFID	radio-frequency identification
RIBA	Royal Institute of British Architects
RICS	Royal Institution of Chartered Surveyors
RMC	ready mix concrete
SAE	Society of Automotive Engineers
SEA	Swedish Energy Agency
SHEQ	Safety, Health, Environment and Quality
SIPOC	Suppliers-Inputs-Process steps-Outputs-Customers
SME	subject matter expert
SPC	statistical process control
SPOT	Smart Personal Object Technology
SSHEQ	Security, Safety, Health, Environment and Quality
T5	Terminal 5
TBT	Tool Box Talk
TPS	Toyota Production System
TQM	Total Quality Management

TRIZ	Theory of Inventive Problem Solving (in Russian)
UAE	United Arab Emirates
UAV	unmanned aerial vehicle
UHI	urban heat island
USACE	U.S. Army Corps of Engineers
UWB	Ultra-wideband
VINNIE	Very Intelligent Neural Network for Insight & Evaluation
VR	Virtual Reality
VSS	voluntary sustainable standards
WBS	work breakdown structure
WLAN	wireless local area network
WUFI	Wärme Und Feuchte Instationär
XR	extended reality

1   Introduction

Every generation thinks they are living in an era of unprecedented change but I believe that in the twenty-first century, we really are starting to see an exponential growth not only in technological capabilities but also in the relationship between humans and technology. Technology is becoming genuinely more autonomous through Artificial Intelligence (AI), based upon continual self-learning from initial rules and algorithms provided by humans. The prospect over the coming decades is that we are heading for designs being developed and engineered by AI, with human interventions reduced to aesthetic choices and construction sites building structures using autonomous vehicles and robots. Operations and facilities management will be overseen by AI, with minimal strategic oversight by humans.

When you look around the seeming chaotic nature of muddy sites with materials scattered around and operatives dodging excavators, it is hard to imagine such a scientific approach to construction but aside from the occasional interjection of media stories about construction technology, there is an undercurrent of sustainable momentum for digital modelling, augmented reality, drones, laser scanning, advanced materials and other technologies quietly appearing in the industry around the globe.

The reasons for such changes are fuelled by the industry's lamentable record in productivity with infamous graphs showing construction flatlining for decades, compared to IT, energy, chemical and manufacturing sectors, such as aerospace and automotive. While virtually all other industries have steadily invested in research and development, the fragmented supply chain, the dog-eat-dog world of contracting and, frankly, the have-their-cake-and-eat-it-too clients have left construction starved of innovation and investment.

A McKinsey report[1] for mega-construction projects of over $1 billion found that the average slippage was twenty months in programme, 80 per cent over budget spend, with a staggering 98 per cent of projects overspent or having taken longer than scheduled. The route to reducing such embarrassing figures is through Digital Quality Management.

A report by Mace[2] demonstrated that if construction's productivity had kept pace with manufacturing over the past twenty years, then the UK would have seen an approximate 3 per cent increase in the country's Gross Domestic

Product (GDP) from each construction worker producing £38/hour of economic activity, compared to £25.50. The tax generated by an additional £100 billion of annual economic activity would have produced an extra £40 billion for the government in tax revenue; equivalent to paying for the following projects, *with £3.3 billion leftover!*

- Terminal 5 at Heathrow (£4.2 billion)
- Crossrail (£14.8 billion)
- Royal Wharf housing development in London(£2.2 billion)
- Queensferry Crossing bridge, north of Edinburgh (£1.35 billion)
- the Mersey Gateway bridge (£540 million)
- Birmingham's Big City Plan (£10 billion)
- the M4 Relief Road around Newport (£1.3 billion)
- Belfast's north-eastern quarter plan (£400 million)
- City of Glasgow College (£228.6 million)
- the A14 upgrade (£1.5 billion)
- Liverpool's triple towers (£250 million)

All those projects could have been paid for if the construction industry had got its act together in the modern age.

A few examples of awful construction project spiralling costs, over-runs and avoidable quality problems are shown in Table 1.1, without going into all the safety and sustainability issues.

In addition, there are examples of the tragic consequences of historic construction failures with root causes of quality management failures:

- 1975, China: Banqaio Dam, 171,000 dead, due to under-designed sluice gates and defective concrete design.
- 1981, the USA: Hyatt Hotel skywalk collapse, 114 dead, due to design change to skywalk steel tie rod connections and drawing version control.
- 1995, South Korea: Sampoong department store collapse, 502 dead, due to substandard concrete mix, original design flaw and air conditioning vibrations.
- 2010, the USA: Deepwater oil rig explosion, eleven dead and environmental catastrophe, due to defective concrete and valve failures.

In the following chapters, I have selected construction projects for a mixture of reasons: iconic status, innovative construction techniques and sheer familiarity and bias upon my part. Those described are not a list of 'greatest hits' and many more could have been chosen.

Managing the quality of design and construction has been a skill set for millennia, dating back to the earliest structures when accepting or rejecting the quality of stone and timber, based on long experience of and expertise in raw materials, was vital for the successful building of a house, burial place or citadel. In order to fully understand the future direction of quality management, it is useful to look back in time to the earliest beginnings of construction to see how

Table 1.1 Examples of recent calamitous construction projects

Project	Initial cost estimate	Final (or latest) cost	Initial completion date	Final (or latest) date	Quality issues inc.
Berlin Brandenburg Airport, Germany	€2bn	€6–7bn	October 2011	2020	• automatic doors installed incorrectly • wiring found to be faulty • escalators too short
Boston's Central Artery/Tunnel Project, US	€2.6 billion	€24 billion	1998	2007	• miscalculation of a tunnel alignment • guardrails lethal in car accidents • corrosion of lighting fixtures • concrete leaks
Olkiluoto 3 nuclear plant, Finland	€3.2 billion	€8.5 billion+	2009	May 2019	• information and communication problems • absence in tender of quality requirements • delays to instrumentation and control system • inadequate welding & poor concrete
Hong Kong-Zhuhai-Macau bridge project*	38.12 billion yuan ($5.5bn)	47 billion yuan ($6.8bn)	2016	2018	• faking of concrete test reports • artificial island movements • seawater leaking into tunnel

Note
* Due to estimates in three different currencies, costs may vary.

our ancestors around the world developed skills and techniques for controlling quality. Over thousands of years, quality control knowledge and experience acquired informed all levels of construction workers and 'managers' and were imbued into their training and supervision.

As quality control steadily gave way to a more strategic approach through, first, quality assurance and then quality management, so the influence of quality professionals grew, until around the turn of the twenty-first century when it seemed that 'quality' was merely a fad to be shunted sideways to make way for Six Sigma, business improvement and business excellence.

The test for me these days is asking which offices are closest to the chief executive's office and usually the response is NOT that of the highest-ranking quality professional. Yet quality management is one key aspect to solving so many challenges that construction management faces in this complex world. If all other disciplines were educated in quality planning, quality control, quality assurance and quality improvement, there would be a significant decrease in overall risk and cost and an increase in quality outcomes of performance.

To achieve a newfound influence within construction businesses, the quality profession needs to raise its game and stop complaining so much. Part of adding value has to be educating others in quality management and how to fundamentally solve problems. Throughout all parts of the construction process, solving challenges is a key skill and yet there is little training and education in appreciating the best tools and the most appropriate circumstances of when to apply those tools. Construction quality professionals need to learn from their counterparts in aerospace and automotive manufacturing and get to grips with quality tools that include not just the basic five Whys, process mapping or the Ishikawa cause and effect diagram, but more sophisticated tools such as 8D investigations, Failure Modes and Effects Analysis (FMEA), Design of Experiments (DoE), Probabilistic Risk Assessment (PRA) and the Quality Function Deployment (QFD) or House of Quality. We can learn powerful lessons from likely causes and options to mitigate risk *before* they occur during design. Quality professionals should have an in-depth knowledge and understanding of these tools in the context of construction to educate and assist other engineering and construction disciplines.

We need to increase the noise levels of how we can help businesses and we need to get into the digital era by becoming its chief protagonists. Digital technology has arrived and is here, whether we like it or not, and quality professionals should be at the forefront of adapting to it and shaping the digital future. Drones are a boon to inspections at height, saving time (and hence money) by creating high resolution images known in three-dimensional space of where precisely they are taken. They provide not just visual representations but also can scan construction components to compare with digital models to check if what has been built is what was designed. Instead of clambering up scaffolding and taking dumb photos, we have the technology at our fingertips to instantly tell the construction manager if the concrete pour, ducting, or brickwork are exactly where they should be and either allow rapid correction or update the design in real time, so that adjustments can be made to associated components. That requires quality professionals to be trained and qualified as drone pilots and pester their managers to purchase drones. We need to then promote the results to demonstrate the improved aspects of quality management, so that senior managers see the discipline in a new light as offering solutions rather than being perceived as complaining.

Digital Quality Management is the new approach to solving problems. It is not about 'digitising' quality management but rather placing information management at the heart of our discipline. Using data and information is how we traditionally offer value to construction through quality management systems, audit reports, inspection and test results, levels of non-conformances and best practices. These limited forms of information, which at best can offer only weak added value to executives making decisions, should be extended much further so that we become the assessors of data quality and use AI to 'audit' in real time all aspects of activities to identify non-conformances and report back quality issues. Suddenly from being one of the backroom boys and girls in construction, the quality professional is front and centre in business intelligence on the performance of a company or project.

The art and science of construction are taking a design and transforming it into a built environment to hand over to the client, which will satisfy their requirements. It takes people capability using machines and processes to turn materials into the final construction. What the client gets is both a physical environment, of concrete, steel, glass, wood, electrical cabling, mechanical plant and other materials, and information, such as an Operations & Maintenance Manual or a Building Information Modelling (BIM) digital representation of the structure.

The problem is that clients, consultants and contractors sometimes forget what information they are delivering in their excitement at creating the physical structure rising out of the ground. Yet without the right information, there can be deleterious long-term effects that can plague the client for years. Where exactly were those cables placed in the wall? Inevitably, repeatedly, time and money are wasted with those design and construction errors forcing higher life-cycle costs and rework to put right. Thinking from the very beginning across the whole project lifecycle as to the information required, reduces the risk to programme, cost and quality.

My definition of quality is 'facilitating the performance guarantee' of the built structure. This goes further than the traditional 'conformance to specification' that many in the construction industry seem to vaguely point to when asked about quality. There is likewise only a superficial acknowledgement of 'customer satisfaction'. Ultimately what construction should be delivering is the required performance of the structure; whether it is the strength in roof beams or aesthetic aspects, say, in an apartment tower block. Quality management cannot guarantee *with certainty* that the construction will meet these client's requirements but it should be the mechanism with which to enable the standards to be met that are agreeable to the client, if conducted professionally and with due diligence. However, quality management is not currently implemented with such fastidiousness and becomes a sop to clients and regulators that the business cares about quality and a punchbag for when things go wrong.

Guarantee is used in the same way that your washing machine has a guarantee. It does not mean that 100 per cent of the time all of manufacturer X's washing machines will work perfectly but it provides a formal assurance that certain performance conditions will be fulfilled, and if something does go wrong that the guarantee can be invoked. It is not in the interests of the supplier of the washing machine for it to break down outside those performance conditions, e.g. it is being installed and used in accordance with manufacturer X's instructions. How does the manufacturer therefore have a high level of confidence that the formal assurance can be given? It places great emphasis on quality management during design and production to meet the requirements.

When the directors of a construction business are sitting opposite the client's representatives across the boardroom table, prior to signing a construction contract, the quality management approach to people, processes, machines, materials and information should be their guarantee that it will meet all the aspects of the contract. However, such executives have typically little appreciation of quality management and fail to understand the fundamental advantages of deploying the necessary resources for effective quality management nor providing the business authority needed for quality professionals to be effective. As their business processes are severely disrupted by new technology, now is the time for a radical re-think and embed Digital Quality Management in the heart of how they deliver projects.

Leveraging the technology available to construction will increase productivity but only if there is a deep understanding of the underlying information management required to use that technology, such as drones, BIM, augmented reality, laser scanning, robotics, 3D printing, blockchains, artificial intelligence, data trusts and a diverse list of other exciting technologies. The quality assurance and quality control techniques that we use today in construction are starting to become outdated and inadequate for managing these technologies and technologically driven processes, so that quality management remains not just relevant but adds more value in the future. However, to accomplish that goal, we need Digital Quality Management with information at the heart of it.

This book is not about a far future, trying to guess what may happen in the construction industry and, in particular, quality management. We live in arguably the greatest time of change with clear patterns emerging of technology and ways of working that will greatly impact on our work. By 2030, many traditional jobs in construction will be gone or radically changed due to the introduction of technology from Artificial Intelligence, robotics, advanced materials, augmented reality, virtual reality, drones, blockchains and many more technologies that are only in research and development (R&D) now or are yet to be developed. However, many new jobs are already being created, such as drone pilots or drone data analysts.

The continual millstones around the neck of construction such as a fragmented supply chain, low profit margins, typically less than progressive clients, impact of outdoor conditions and other 'excuses' will be removed, as more offsite manufacture and assembly take place and data trusts and BIM reinforce collaboration and data sharing to improve effectiveness and efficiencies. Knowledge will be better retained inside industry data trusts and combined with intuitive knowledge management and quality management systems that allow subject matter experts, in all fields, to easily access and assist with driving business innovation. So, if the traditional problems of an inefficient, bureaucratic industry start to disappear, then more investment will flow into the industry, increasing the speed of innovation and improvement. Unless leading decision-makers in the larger construction contracting businesses adapt, they will find that their business models are likely to become obsolete.

The solutions to the ingrained problems are gathering momentum and as they become commonplace and evolve efficient processes, so the early adopters will find they have a significant competitive advantage.

The key issue is research and development into new and existing technology to exploit its potential. With traditionally low margins, individual business R&D in construction is tiny compared to other industries. This needs to change. Construction businesses that now are working in joint ventures should collaboratively invest in innovation. Such improved collaboration will deliver greater profit margins in the long term.

Many of the old construction practices will remain, especially for smaller projects, as some companies may be able to deliver a profit with traditional niche skills in the same way that expert stonemasons today are required to maintain historical buildings. However, all companies run the risk of becoming redundant if they continue to use inefficient processes and ignore new technology.

Quality professionals need to be educated and continually educate themselves in the changes and trends in technology so that they can build such disruption and significant impacts into strategic policies in their thinking. They do not need to become expert programmers but understanding the basic premise and reasoning behind programming will enhance their ability to converse with their chief information officers (CIOs) and appreciate how to exploit such technologies for the greater good of their people in the business. By being digitally capable, quality professionals can become better customers of information management and IT departments, knowing what it can and cannot do and ensuring that the fundamental philosophy of lean construction and quality management is the underpinning driver for decision-making rather than simply accepting suggestions from IT professionals.

Where technology has been introduced into other disciplines, such as health and safety, environmental management and security management, we should be assessing if these applications can be applied to quality management. I have endeavoured to explore how technology has been exploited and highlighted that it may be applied for quality management but much more research and experimentation are needed to identify practical applications for training, communications, auditing and management systems. Quality professionals need to be brave and inquisitive to recognise deficiencies in our discipline and figure out if technology can supply solutions.

Bots may provide e-learning modules of aspects of problem solving learning that are far more enjoyable and intuitive than costly classroom teaching, which are vulnerable to annual budget-cutting exercises and require minimum numbers to be viable. Those bots need to be created and formulated by expert quality professionals, so that the best knowledge is being imparted. There is a responsibility on quality professionals to champion such solutions and help design them so that the quality knowledge of our profession is embedded into AI and helps shape it for future generations.

We need to take the best practices and principles of quality management and create a new approach that delivers a demonstrably better built environment by leveraging technology and innovation from around the globe. In this way, quality management will be in the vanguard of optimising the results of information and knowledge management and become a bedrock in construction businesses.

Above all, quality professionals should keep their 'eyes on the prize' of customer satisfaction and safety. To achieve that end, they must act and be perceived to act independently of all the daily pressures heaped on them to agree that there are not any 'real quality' issues. Senior managers often do not want to hear about the quality problems on top of cost and programme problems but they need to listen to them. If they don't hear it from us, then who is going to tell them? At the same time, however, as setting out the measurable problems, we need to set out the measurable solutions.

If we do not adapt rapidly, then our discipline may well die as management systems' standards, driven by generic factors such as risk, move closer together. There may come a time when having separate disciplines for health and safety, environment, quality, business continuity and information security become obsolete, either through multi-skilled operatives developing sufficient 'coal face' knowledge and/or through a powerful AI that can assist them on risk management to the point that a full-time quality professional is as rare as a clerk of works.

Notes

1 Changali, S., Mohammad, A., and van Nieuwland, M., 'The construction productivity imperative' (McKinsey Global Institute, July 2015). Retrieved from www.mckinsey.com/industries/capital-projects-and-infrastructure/our-insights/the-construction-productivity-imperative (accessed 14 July 2015).
2 Mace, 'Construction productivity: The size of the prize'. Retrieved from www.macegroup.com/perspectives/180125-construction-productivity-the-size-of-the-prize (accessed 24 January 2018).

2 Ancient world construction quality, Palaeolithic–CE 500

For thousands of years, humans have created and built structures in which to live, work, pray and bury their dead. We may not always see the remains of some of the earliest buildings but that yearning to use materials around us to create spaces that are ours, has been a passion and an obsession since we left the caves, animal skin tents and pit-houses.

There have been 'inspectors' and expert 'supervisors' who undertook quality control-type duties throughout history from the earliest Palaeolithic times when humans started to build structures. They were the time-served craftspeople and artisans who knew what quality looked, sounded and felt like. The oldest structures (or their remains) that are still visible attest to the quality of the building materials, techniques and the skills of their builders including:

- the oldest person-made ruins on Earth; a stone wall at Theopetra Cave in Greece, c.23,000 BCE
- the Göbekli Tepe sanctuary or temple ruins in Turkey, 9500 BCE
- the Jericho Wall and Tower in the Palestinian Territories, c.8000 BCE
- Çatalhöyük village in Turkey, c.7500 BCE
- the Khirokitia Neolithic settlement in Cyprus, c.7000 BCE
- the Barnenez passage grave in France, 4800 BCE
- the Knap of Howar house in Scotland, 3700 BCE
- Sechin Bajo Plaza in Peru, 3500 BCE
- Shahr-e Sūkhté village in Iran, c.3200 BCE
- Stonehenge standing stones in England, started in 3100 BCE (Figure 2.1)
- the Maikop kurgans tombs in Russia, 3000 BCE
- Sadd al-Karfara, the 'Dam of the Pagans' in Egypt, c.2750 BCE
- Dholavira city in India, c.2650 BCE
- the Great Pyramid of Giza in Egypt, constructed in 2540 BCE
- the Palace of Knossos in Greece, 2000 BCE

The astonishing accuracy of some ancient buildings is evidence not only of an architect's brilliance in imagination and design but also the construction managers' practical construction expertise and their precise quality control. The Great Pyramid of Giza, dating from 2540 BCE, comprises 2.3 million stone blocks

Figure 2.1 Stonehenge, England, 3100 BCE.

(comparable in volume to the Hoover Dam, completed in 1936) weighing between 2.5 and eighty tons, with a work rate of laying an average-sized block every three minutes throughout a ten-hour working day.[1] The accredited architect, Hemiunu, designed a structure that was built without wheels, compass or pulleys with the most basic of tools of copper chisels and saws and stone hammers. The 5.3 hectares at its base were levelled to within 2 cm accuracy. Whether a coincidence or by design is unknown, but the area of one triangular face is equal to the square of its height. The white limestone casing blocks were finished so precisely that they did not require mortar and the joints were within 0.5 millimetres in accuracy.[2] That is construction quality par excellence and is as good as or better than modern buildings and infrastructure with all their know-how, machinery and technology. However, a key difference is the use of slaves on many of these construction projects, whose individual names may never be known but who collectively should be seen as deserving recognition for their skills, knowledge and hard labour that contributed to incredible engineering feats, albeit under enforced conditions.

While many were paid for their work in labouring to construct magnificent structures over the millennia, hundreds of thousands of slaves were also used and abused. They had no choice but to follow instructions in lifting, carrying, cutting and crafting edifices in stone, timber, brick, mud and rock. Most great architects in history will have made use of slaves, who would have offered or been asked on their expertise, gleaned through years of enforced service. These slaves remain nameless and yet have also solved building challenges and positively affected and improved the marvels of the built world. We remember those slaves and acknowledge that they suffered so much in the constructed creations. One such slave was Abram Petrovich Gannibal (1696–1781), son of a tribal prince who was kidnapped from his native Cameroon and after enslavement by the Turks, was passed to Russia and the court of Peter the Great. The Tsar adopted him and he had an incredible career as a military engineer and mathematician, extending the Ladoga Canal and building a number of Russian fortresses, rising to the rank of General. His great-grandson was Alexander Pushkin, the famous Russian poet.

The fragility of the writing materials through the ages from silk and bamboo in ancient China to papyrus, tree bark and parchment in the ancient Indus Valley civilisation, Egyptian, Greek and Roman times has meant that many texts on methods, inspection, testing and standards of construction have been lost. The texts that have survived demonstrate that construction standards have been developed and mastered through the millennia. Those that have been written down suggest that they were orally passed from generation to generation over thousands of years before.

In India, Hindu temples started to appear in the Indus-Saraswati era or Harappan Civilization (2600–2000 BCE) as open-air places to worship but these were enclosed in the Mahajanapada period (700 BCE) in the form of thatched huts. By 200 BCE, multi-storeyed temples were being constructed and, by the late fifth century, stone temples were built without mortar. Many important texts on architecture and construction, the *Vedas*, the *Brahamanas*, the *Upanishads* and the *Bhagvad Gita*, have details on measurement techniques, materials and specifications with the aim of establishing the relationship in these hallowed places between people and gods.

The layout of temples is provided in the *Shilpa Shastras* and *Vastu Shastras* and two of the most important texts in Southern India are the *Mayamata* and *Mansara*, which lay down precise rules on the architecture and iconography. For example, the quality of stone was based upon its colour, hardness and the touch of it against the fingers. By the seventh century, the codes of building temples were firmly established and adhered to. Similar to the Chinese Feng Shui, an aesthetic quality for the temples is blended with an appreciation of nature in the *Vastu Shastras*.[3]

The Babylonian code of law of ancient Mesopotamia known as the Code of Hammurabi, dates back to about 1754 BCE and makes clear the importance of building standards, with the added incentive of facing the death penalty if the builders get it seriously wrong: 'If a builder build a house for someone, and does not construct it properly, and the house which he built fall in and kill its owner, then that builder shall be put to death.'[4]

The Temple of Solomon in Jerusalem, completed in 950 BCE (before being destroyed in 586 BCE), had stone that was apparently quarried and finished offsite (what we call Design for Manufacture and Assembly, DfMA, these days) before being assembled by Phoenician artisans. No mortar was used as the stones were cut with such precision. This demonstrates a clear building quality control process to ensure that only stone that met the strict, quality control specifications was transported ready for assembly. 'In building the temple, only blocks dressed at the quarry were used, and no hammer, chisel or any other iron tool was heard at the temple site while it was being built.'[5]

The Great Wall of China was started as a string of individual walls around the eighth century BCE. King Xuan instructed General Nan Zhong to keep out the nomadic tribe of the Xianyun. Over the centuries, stretches of walls were built to protect the northern borders with the Great Wall of Qi in 685 BCE a notable construction of 600 km in length. In 212 BCE, Qin Shi Huang, the first emperor of China, commanded the building of a series of walls that ran from Gansu to the coast in Manchuria. The Qin records suggest that between 300,000 and 500,000 soldiers with another 400,000–500,000 peasants were conscripted to build the wall. Under the Han, Northern Qi, Sui and Ming Dynasties, the wall was further extended with forts, embankments and beacon towers through to the seventeenth century.

The Great Wall measures an astonishing 21,196 km in total, through all its branches and natural defences including hills and rivers. In its earliest phases, rammed earth, stones and wood were used but, under the Ming Dynasty, sections were constructed with kiln-fired, standard-sized bricks. Many of these bricks are stamped with names and dates of the production workshop as part of a quality control process to trace back any unacceptable batches.

The Old Testament in the Bible makes a number of references to building quality. The Book of Kings, written perhaps around 600 BCE, says:

> All these structures, from the outside to the great courtyard and from foundation to eaves, were made of blocks of high-grade stone cut to size and smoothed on their inner and outer faces. The foundations were laid with large stones of good quality, some measuring ten cubits [4.5 metres] and some eight [3.6 metres]. Above were high-grade stones, cut to size, and cedar beams.[6]

Examples are found around the globe of exquisitely cut rock with near-perfect planes. The ancient Greeks took their building skills with them on the travels as they explored and settled in new lands. At Saint Blaise, near Bouches-du-Rhône in southern France, the archaeological site, that dates back to 4700 BCE, has revealed such exact Greek stonemasonry from 650 BCE, overlaying earlier ramparts, shown in Figure 2.2.

In Ancient Greece, Plutarch, writing about Pericles' great reconstruction plan after the wars between Greece and Persia, ended with the Peace of Callais Treaty in 449 BCE, set out the new building works for the Acropolis:

> [H]e [Pericles] boldly suggested to the people projects for great constructions, and designs for works which would call many arts into play and involve long periods of time, in order that the stay-at-homes, no whit less than the sailors and sentinels and soldiers, might have a pretext for getting

Figure 2.2 Walls at Saint Blaise, near Bouches-du-Rhône, built 650 to 625 BCE.

a beneficial share of the public wealth. The materials to be used were stone, bronze, ivory, gold, ebony, and cypress-wood; the arts which should elaborate and work up these materials were those of carpenter, moulder, bronze-smith, stone-cutter, dyer, worker in gold and ivory, painter, embroiderer, embosser, to say nothing of the forwarders and furnishers of the material, such as factors, sailors and pilots by sea, and, by land, wagon-makers, trainers of yoked beasts, and drivers. There were also rope-makers, weavers, leather-workers, road-builders, and miners.[7]

Architects in ancient Greece, when designing temples, also took hands-on roles on site, as project managers. In 449 BCE, an Athenian law set out that the Temple of Athena Nike, on the Acropolis overlooking Athens, must be built to the exacting description laid down by the architect Kallikrates. Designed in a four-column (tetrastyle) Ionic construction (the North Portico of the White House in the USA shown in Chapter 4, Figure 4.1, is an example of the ongoing love affair with this type of architecture), the rear and front façades have a colonnaded portico (amphiprostyle). The law also stated:

and that the sanctuary be provided with gates in whatever way Kallikrates may specify; and the official sellers are to place the contract within the prytany of Leontis; the priestess is to receive fifty drachmas and ten to receive the back legs and skins of the public sacrifices; and that a temple be built in whatever way Kallikrates may specify and a stone altar. Hestiaios proposed: that three men be selected fifteen from the Council; and they shall make the specifications with Kallikrates and ... in accordance with [the contracts].[8]

Kallikrates was also the joint architect with Ictinus of the Parthenon (see Figure 2.3) on the Acropolis, begun in 447 BCE and widely regarded as one of the finest examples of Greek architecture, breaking traditional rules of design with an extra interior room, number of columns different from the previous standard formulae and with a frieze around the exterior, celebrating the Greek victory over the Persian Empire. It is likely that Pericles designated Phidias as the first architect in overall charge of the collection of buildings on the Acropolis, which includes the Parthenon, Propylaea (the gateway), the Erechtheion (a temple) and the Temple of Athena Nike.[9] Phidias was a renowned sculptor as well as an architect, designing the Parthenon's Marbles (commonly known as the Elgin Marbles, now housed in the British Museum, and understandably claimed by the government of Greece) and later he created the colossal Statue of Zeus at Olympia, one of the Seven Wonders of the Ancient World, in 435 BCE.

The meticulous approach to quality control is highlighted by the Greek procedure for checking that blocks of stone had been cut to exceptionally smooth

Figure 2.3 The Parthenon, on the Acropolis above Athens, Greece, finished in 432 BCE.
Source: pexels.com.

planes so that they would fit together. A red pigment, vermilion, was applied to a *kanon* stone (meaning 'measure' or 'rule') that had been sculptured to near perfect flatness. The *kanon* would then be placed on the surface of the quarried stone block and then removed. If there were any spots on the surface without vermilion, then it showed that the stone block was not sufficiently smooth and flat and the mason would be instructed to cut away the uneven parts.

While the Egyptians had used many slaves to move large carved stones, the Greeks evolved methods of more easily handled stone blocks. Temple columns, for example, were usually cut into convenient drum components that were then bolted together using wooden pins that slotted into prismatic-shaped *empolia*,[10] made from cypress wood, which in turn were set into sockets cut out of each stone drum. The pin technology helped to accurately centre the drums and by turning the top drum each time a few degrees, it achieved greater purchase, without the need for mortar.

Greek tools, in addition to the *kanon*, also included the *diabetis*, a surveying spirit level, which was made from wood in the shape of the letter A with a cord hung from the apex with a stone at the end. If the stone inclined too far in one direction or another, the stone block could be seen to be not flat. Excavations at the Temple of Apollo at Didyma have revealed how the stonemasonry technique of *entasis* was crafted. This tapering of the Ionic columns for aesthetic effect was achieved through numbering each column drum on how much material to cut and where it was to be cut.[11]

The Greeks were masters of architectural building proportions. The proportion of length to width of buildings was centred on the preferred ratio of 1:1.6 (or more exactly 1.61803), the so-called 'golden section', referred to in Euclid (see Chapter 7 on data), based upon Phi *Φ*, and appearing in the numerical series called the Fibonacci sequence after the twelfth-century mathematician. Instinctively, many people seem to prefer this ratio and it may have been derived from many years of building design that seek to harmonise nature, the human appreciation for aesthetics and architecture. The number of petals in a flower usually follows the golden ratio and each petal is placed at 0.61803 per rotation, to maximise sunlight exposure. The human face, seed heads, pinecones and shells typically follow the golden ratio. Even the Parthenon, built 150 years before the golden ratio was formally recorded, seems to follow it in some parts of its design.[12] While never an overriding design concept, it does seem to appear to be hard-wired into our aesthetic preferences.

In 403 BCE, under the Zhou dynasty in China, the *Kao gong Ji* (Record of Inspecting Public Works), covered the construction of palaces, cities, moats, among other crafts and stated: 'Heaven having time, earth having energy, material having beauty, work having skill, add these four and the result is quality.'[13]

Roman roads were not only a demonstration of transportation efficiency in moving armies and logistics around the Roman Empire but also a vital information network. One of the most famous of the Roman roads, the Via Appia, started in 312 BCE, ran for 132 miles between Rome and Capua, before being extended a further 220 miles to eventually reach Brindisi in 244 BCE.

The poet Statius extols the virtues of Emperor Domitian's road building patronage in CE 95:

> The first labour was to mark out trenches,
> Carve out the sides, and by deep excavation
> Remove the earth inside. Then they filled
> The empty trenches with other matter,
> And prepared a base for the raised spine,
> So the soil was firm, lest an unstable floor
> Make a shifting bed for the paving stones;
> Then laid the road with close-set blocks
> All round, wedges densely interspersed.
> O what a host of hands work together![14]

One of the Roman stars of the construction quality pantheon was Marcus Vitruvius Pollo (born *c.*80–70 BCE and died some time after *c.*15 BCE), commonly known just as Vitruvius (and immortalised by Leonardo da Vinci in his drawing of a naked man with outstretched arms in a circle and a square as the 'Vitruvian Man'), who published ten books,[15] covering subjects of 'points of excellence' for construction materials, architectural design, machines and the physics of building.

A test for the cleanliness of sand that is good enough to be mixed into mortar was to 'throw some sand upon a white garment and then shake it out; if the garment is not soiled and no dirt adheres to it, the sand is suitable'.

He despaired at wattle and daub on fire safety grounds; it 'is the disaster that it may cause, for it is made to catch fire'. Vitruvius spelled out the importance of allowing the sap of newly felled timber to drip out 'drop by drop' to avoid the timber dying 'in a mass of decay, thus spoiling the quality of the timber'.

He praised the Roman concrete called pozzolana, after the type of volcanic ash mixed with lime and rubble, as it 'not only lends strength to buildings' but 'even when piers of it are constructed in the sea, they set hard under water'.

Vitruvius dedicated his *De Architectura* to the Emperor Augustus and it is worth pointing out that many of the quality management techniques devised and refined from the Greeks and the Etruscans to the Romans were by skilled contractors and craftspeople but their expertise has been not recognised and it was rich Roman citizens and their patrons whose names are written into history.

Sextus Julius Frontinus (CE 30–103 or 104) was a Roman magistrate, praetor, consul, Governor of Britain between CE 74 and 78 and a writer on military matters and civil engineering. He wrote at length about the aqueducts of Rome, *De Aquaeductu Urbis Romae* and parts of his treatise on surveying have also survived.

He mentioned the accuracy of building and maintaining aqueducts;

> The channels of the aqueducts without [outside] the city must be carefully examined, one after the other, to review the granted quantities [of water];

the same must be done in case of the delivery tanks and fountains, that the water may flow without interruption, day and night....[16]

Frontinus also described two of the key Roman surveying instruments: the *groma* and *chorobates*, which have since been recreated in modern times and proven to be accurate in setting out.

The *groma* was a rod planted into the ground with a horizontal cross-piece affixed at the top of the rod with a plumb line dropping down from each end of the cross-piece. An assistant would be sent some 100 paces in the direction of the intended line and hold a rod vertically. The surveyor, by lining up two plumb lines, could then instruct the assistant to move left or right until the two plumb lines and the assistant's rod were aligned. Additional lines could be plotted at right angles by sighting through the other parts of the cross-piece.

The *chorobates* was a type of very large, timber spirit level, some six metres in length on legs to the height of the surveyor, with plumb lines at each end and a '*canalis*' groove in the middle, filled with water to show when the apparatus was level. This could then be used to eye-in a measuring rod down the slope of a hillside, to calculate the vertical height.

The *dioptra* was also used by Roman surveyors like a theodolite, especially for steep slopes. Described by Heron of Alexandria (CE 10–70), the *dioptra* was a small metal disc fitted in the top of a rod grooved with angles. By using screws, the disc could be made horizontal through observing two water levels and using a rotating bar with sights, the surveyor rotated the disc to calculate the angle between two objects and used triangulation to calculate distance.

Vitruvius credited Caius Sergius Orata as the inventor of the hypocaust, the underground heating system. In 95 BCE, Orata was an entrepreneur and noted the Roman fashion for eating oysters, so he introduced oysters into Lake Fucino and the volcanic minerals in the water gave them a distinctive taste. Orata became very wealthy but any cold winters would kill off his oyster stocks. To prevent this, he devised a water basin to hold them, with an underground ducting system that led to a fireplace that kept the temperature just right. The design of the hypocaust was said to have been copied for use in Roman buildings.

The Buddhist Pail canonical works called *Tipitaka*, from around the CE first century mention the role of the 'Maintenance and Construction Official' and their duties,[17] metal standards,[18] and the strength and safety of constructing walls.[19]

Building collapses conversely show when quality control has catastrophically failed. In CE 27, a wooden amphitheatre in Fidena, built by a freedman called Attilius, collapsed after some 50,000 spectators had entered for games under the Emperor Tiberius.

men and women of every age crowding to the place because it was near Rome. And so the calamity was all the more fatal. The building was densely crowded; then came a violent shock, as it fell inwards or spread outwards, precipitating and burying an immense multitude which was intently gazing on the show or standing round.

It was estimated that all 50,000 people were killed or maimed. As Tacitus, the Roman historian wrote, by decree of the Senate no one was to exhibit a show of gladiators, whose fortune fell short of 400,000 sesterces, and no amphitheatre was to be erected except on a foundation, the solidity of which had been examined.[20] Likewise part of the upper spectator decks at the sporting arena, Circus Maximus collapsed in CE 140, killing 1,112 people. It could hold around 250,000 spectators and the lower decks for the aristocratic families were made of stone but the upper decks were typically wooden frames.

Lake Fucino was the third largest freshwater lake in Italy but had a reputation for causing flooding in the surrounding areas and being infested with malaria. Emperor Claudius, looking for new fertile land for Rome, revived an earlier plan by Julius Caesar to drain the lake. So, in CE 41, an eleven-year construction project began, using 30,000 slaves to dig a three mile underground tunnel and partial culverts to create a new five-mile outlet to the River Livi. At the time (and until the eighteenth century) it was the longest tunnel in the world. The project was only a partial success and Emperor Hadrian commissioned excavation work to deepen the canal in places. With the collapse of the Roman Empire, the canal silted up but the plan was again revived in 1862 as a major project by Prince Alessandro Torlonia, with the lake completely drained by Swiss engineer, Jean François Mayor de Montricher, and now the area is very fertile farming land.

The Colosseum in Rome is considered one the 'modern' seven wonders of the world. Started in CE 72 and taking just eight years to complete, it is estimated that over 100,000 cubic metres of travertine stone and 300 tons of iron clamps[21] were used to build the main structure, with outside walls rising to forty-eight metres. Following the defeat of the Jewish Revolt of CE 71, it is likely that thousands of Jews were used as slaves in the construction.

The scarcity of professionals in the construction industry seem to have been a recurrent theme in history. When Pliny the Younger wrote to the Roman Emperor Trajan in CE 111 asking for a *mensor* (surveyor) to be sent to survey public works in northern Turkey with the intention of recovering money from contractors, the emperor responded by complaining about the lack of surveyors available: 'I have scarcely enough surveyors for the public works which are in progress at Rome or the immediate district....'[22]

In a later letter that same year, Pliny begged Trajan for further assistance in completing a theatre and public baths:

> I beg you not only for the sake of the theatre, but also for these baths, to send an architect to see which is the better course to adopt, either, after the money which has already been expended, to finish by hook or by crook the works as they have been begun, or to repair them where they seem to require it, or if necessary change the sites entirely, lest in our anxiety to

save the money already disbursed we should lay out the remaining sums with just as poor results.[23]

And again, Trajan firmly declined:

There must be plenty of architects to advise you, for there is no province which is without some men of experience and skill in that profession, and remember again that it does not save time to send one from Rome, when so many of our architects come to Rome from Greece.

Finally, when Pliny proposed to connect Lake Sapanca to the sea by way of excavating a canal, the emperor relented:

You will be able to obtain a surveyor from Calpurnius Macer, [Senator Publius Calpurnius Macer who governed Lower Moesia, which is now an area between Bulgaria and Romania] and I will also send you someone who is an expert in that class of work.[24]

For whatever reason, the canal was never built. Perhaps the surveyor was re-directed to work on the theatre and public baths or perhaps Pliny gave up hope of finding a skilled surveyor.

The Pantheon, shown in Figure 2.4, is another incredible Roman building built between CE 118 and 125. The Romans had a great understanding of the impact of heavy materials on structures and built the dome with progressive lighter materials as it moved towards the top opening. The initial level con-sisted of traditional travertine stone, then a mixture of travertine and tufa. The next level consisted of tufa and brick, then all brick and finally pumice on the ceiling of the dome. The building was given additional protection from the typical ravages of time when it was converted from a temple into the Church of Santa Maria and Martyrs in CE 609 but, regardless, it is extraordinary that it still remains the largest unreinforced concrete dome in the world.

Teotihuacán, north-east of Mexico City, is home to stunning pre-Aztec pyr-amids built around CE 100 and believed to be the largest city in the Western Hemisphere from that time through to 1400, with 100,000–200,000 inhabitants at its peak.[25] The Aztec name of Teotihuacán means 'the place where the gods were created', since the Aztecs thought that others had built the city. When Hernán Cortes, the Spanish conquistador, asked them, they replied, 'We were not the builders of Teotihuacán, this city was built by the Quinanatzin, a race of giants who came from the heavens in the times of the second sun.' Whoever did build the city had impressive architectural and construction quality experience and know-how.

Along the 1.7 km Avenue of the Dead, there are three pyramid types. The Pyramid of the Sun stands sixty-three metres tall with its square base measuring 225 metres long on each side. It is thought to be one of the largest structures created in the pre-Columbian New World and was completed around CE 200.

Figure 2.4 The Pantheon in Rome, completed in CE 125.
Source: pexels.com.

The Arch of Constantine, dedicated in CE 315, is a turning point away from the classical Greek design that the Romans had imitated. It was built to commemorate the victory of Emperor Constantine I at the Battle of the Milvian Bridge in CE 312 where he defeated Emperor Maxentius, who had occupied Rome. The arch was either constructed in rapid time, partly due to the façades recycling reliefs and sculptures from other monuments, or it was a pre-existing arch that was simply refaced. The core of the arch was travertine stone with marble façade and, standing at 21 metres high, 25.9 metres wide and 7.9 metres deep, was the largest triumphal arch constructed in Rome. This arch typifies the reuse of so much building materials through time. It was far easier to take existing worked stone that met quality standards and move to a new building site than quarry rock from miles away and have to transport it.

The quality management challenges in those ancient times were stunning. Aside from the perennial problem of inclement weather messing up schedules and plans, wild animals and unfriendly tribes declaring war, there was the considerable difficulties of acquiring materials from vast distances. The Stonehenge builders, 5,000 years ago, transported the eighty inner blue stones from the Presili mountains 160 miles away by raft or by dragging them, with each one weighing up to three tons, and the outer Sarsen stones weighing up to thirty tons from twenty miles away, probably by dragging them on a wooden sled.[26] How long did it take to reconnoitre the long distances to find the stones and then have the ingenuity to quarry out and work out a method to transport them, all the while feeding the workforce and giving them incentives to co-operate? How many stones shattered or broke from quarrying or on the journey? With no metal available, the only tools would have been hammerstones made from rock that would have not only shaped the stones but also created mortise and tenon joints for the pi-shaped trilithons. It may have taken ten years of stonemasons chipping away, requiring hundreds of hammerstones. The labourers, craftspeople, planners and designers together with the logistical experts of feeding and housing thousands of people were intelligent, thoughtful and expert builders. The quality of building of Stonehenge and other ancient structures has withstood the trials of the wind, rain, cold and sun, theft of materials and vandalism, to inspire us through the millennia.

Notes

1 Bartlett, C., *The Design of The Great Pyramid of Khufu*. Retrieved from https://link. springer.com/content/pdf/10.1007%2Fs00004-014-0193-9.pdf (accessed 14 May 2014).
2 Smith, C.B., *How the Great Pyramid Was Built* (London: Penguin Random House, 2018).
3 *Vastu Shastras*. Retrieved from www.vastushastraguru.com
4 The Lillian Goldman Law Library, *Code of Hammurabi*. Trans. King, L.W. (2008). Retrieved from http://avalon.law.yale.edu/ancient/hamframe.asp
5 The Bible, 1 Kings 6:7. New International Version.
6 The Bible, 1 Kings 7:9–11. New International Version.

7 Plutarch, *Pericles*, Trans. Dryden, J. (1996), Chapter 12. Retrieved from https://people.ucalgary.ca/~vandersp/Courses/texts/plutarch/plutperi.html#XII
8 Temple of Athena Nike inscription. In *Inscriptiones Graecae* IG I3 35. Trans. Lambert, S., Blok, J. and Osborne, R. (2013). Retrieved from www.atticinscriptions.com/inscription/IGI3/35
9 *Encyclopaedia Britannica*, 'Parthenon'. Retrieved from www.britannica.com/topic/Parthenon.
10 Papadopoulos, K. and Vintzileou, E., 'The new "poles and empolia" for the columns of the ancient Greek temple of Apollo Epikourios'. (2013). Retrieved from www.bh2013.polimi.it/papers/bh2013_paper_229.pdf
11 Rehm, A., *Didyma II: Die Inschriften* (Berlin, 1958–68), No. 48.
12 Meisner, G., 'The Parthenon and Phi, the Golden Ratio'. Retrieved from www.goldennumber.net/parthenon-phi-golden-ratio/ (accessed 20 January 2013).
13 Xiyi., L., (c.1235) *Kao gong ji* (New York: Routledge, trans. 2013).
14 Publius Papinius Statius, *Silvae*, Book IV: 3, the Via Domitiana (Cambridge, MA: Loeb, 2003).
15 Vitruvius, *Ten Books of Architecture* (Cambridge: Cambridge University Press, 2001).
16 *De Aquaeductu Urbis Romae*, Para. 103 Trans. C. Herschel (1899). Retrieved from https://watershed.ucdavis.edu/shed/lund/ftp/Frontinus-Hershcel.pdf.
17 BDK Daizokyo Text Database. *Pāli Tripitaka*. B2025, Chapter 6, p. 135, *The Baizhang Zen Monastic Regulations*. (Trans. Shohei Ichimura). Retrieved from http://21dzk.l.u-tokyo.ac.jp/BDK/bdk_search.php?skey=construction&strct=1&kwcs=50&lim=50
18 BDK Daizokyo Text Database. *Pāli Tripitaka*. B2025, Chapter 5, p. 68.
19 BDK Daizokyo Text Database. *Pāli Tripitaka*. B0192, Chapter 22, p. 157. *A Biography of Sakyamuni*. Trans. C. Willemen. Retrieved from http://21dzk.l.u-tokyo.ac.jp/BDK/bdk_search.php?skey=wall&strct=1&kwcs=50&lim=50
20 Tacitus, *The Annals*, Book 4, p. 62. Retrieved from https://en.wikisource.org/wiki/The_Annals_(Tacitus)/Book_4#62. Translation based on A.J. Church and W.J. Brodribb (1876).
21 See the-colosseum.net. Extra data from Cozzo, G., *Il Colosseo* (Rome: Palombi, 1971). Retrieved from www.the-colosseum.net/architecture/la_costruzione_en.htm
22 Pliny the Younger, *Letters*. Trans. J.B. Firth (1900), Book 10, Letter 18. Retrieved from www.attalus.org/old/pliny10a.html
23 Ibid., Book 10, Letter 39.
24 Ibid., Book 10, Letter 42.
25 The Metropolitan Museum of Art. Retrieved from www.metmuseum.org/toah/hd/teot/hd_teot.htm (accessed October 2001).
26 Mosher, D., 'It's official: Stonehenge stones were moved 160 miles'. *National Geographic Magazine*, 24 December 2011. Retrieved from https://news.nationalgeographic.com/news/2011/12/111222-stonehenge-bluestones-wales-match-glacier-ixer-ancient-science/

3 The Middle Ages and the Discovery Age in construction quality, 500–1800

Time and again as we travel through written history we come across those men and women whose achievements in the built environment were created using quality management techniques and knowledge that have resolutely lasted through centuries. In Europe, there has been a misunderstanding or even incorrect education that gives the impression that once the Romans departed, then the Dark Ages brought nothing but destruction and decay of Roman buildings. In fact, the Anglo-Saxon culture had sophisticated construction methods but they worked mainly with timber and there are few remains left to showcase such expertise. The oldest timber building in the UK, and one of the oldest in Europe, is the Church of St Andrew, in Greensted-juxta-Ongar in Essex, with the timber framework dated to between 1063 and 1108, built on the location of an older structure dating to at least the seventh century.

The largest building in the world at the time of its construction in CE 537, the Hagia Sophia in Istanbul, Turkey (Figure 3.1) was first a Greek Orthodox Christian patriarchal cathedral, then an Ottoman imperial mosque and now is a museum (Ayasofya Müzesi). It is seen as the epitome of Byzantine architecture. Designed by Anthemius and Isidore as architects, and taking just five years to build (and replacing two previous churches that had been destroyed), it was almost inevitable that the building haste caused the dome to collapse in CE 558, following an earthquake. Its replacement was raised an additional six metres in height to measure thirty-two metres wide and extend 55.6 metres from ground level (the Statue of Liberty would fit inside it). Built from brick and mortar, the new dome used the first large-scale incorporation of four pendentives; spherical triangles of vaulting, to distribute the load from the round areas of the dome to the square-shaped walls.

In India, Vārāhamihira was a brilliant astronomer and mathematician who wrote about many matters. In his *Brihat Samhita* text, written in the mid-sixth century, he wrote about the quality of plaster needed inside temples and that it could be made from a paste of 'adamantine glue': 'Take unripe Tinduka fruits, unripe wood apples, flowers of silk cotton, seeds of Sallaki, the bark of Dhanvana and Vacha, boil them all in a Drone of water and reduce it to an eighth of its volume.' Then add other ingredients and the end result is:' When this glue being heated, is applied to temples, mansions, windows, Siva's emblems, idols,

Figure 3.1 Hagia Sophia in Istanbul, Turkey, constructed in CE 537.
Source: pixabay.com.

walls and wells, it will last for a crore of years.'[1] Since a 'crore' is ten million, this suggests he was totally convinced of the quality of this glue!

Ancient Hindu texts also describe various building material techniques for quality control. Microscopic defects in stones, invisible to the human eye, were detected by applying paints made from herbs so that the defects became visible. The age of stones was determined by hitting with a hammer and categorised as 'child', 'young' or 'old' with only the 'young' stones that sounded like a bell when hit, being used for construction. Likewise, the suitability of stones for building use could be identified from the uniformity of their colour and the presence of lines and spots.[2]

The oldest bridge still standing was built between CE 595–605, known as the Anji Bridge, literally meaning the 'Safe Crossing Bridge'. It was designed by the stonemason Li Chun during the Sui dynasty and spanned fifty metres over the Xiao River in Zhao county of Hebei Province of China. Constructed from quarried limestone slabs that were shaped into curves and joined with iron dovetails to allow for movement. It is nine metres wide with an open-stranded design that reduced weight but maintained the strength of the arch. An inscription added seventy years after its construction testified to Li Chun's achievement: 'Such a master-work could never have been achieved if this man had not applied his genius to the building of a work which would last for centuries to come.'[3]

The Quran, written within twenty years of the Prophet's death in CE 632 makes reference to building standards as a simile for strength of soldiers on the field of battle, with the words 'battlefield formations firm as an unbreakable concrete wall'[4] and when:

> Pharaoh said, 'O eminent ones, I have not known you to have a god other than me. Then ignite for me, O Haman, [a fire] upon the clay and make for me a tower that I may look at the God of Moses.'[5]

Meaning that the bricks were to be baked to a sufficient hardness that could create a tower. Without any type of pyrometer and thermocouples to measure temperature, it took the skill of a professional brick-maker to monitor the colour of the kiln fire and judge the right temperature to produce the required standard of bricks. This type of expertise in brick-making goes back much further in history, to perhaps around 4000 BCE in Mesopotamia and extended forwards to the Romans and after an hiatus across Europe, reappeared in the Middle Ages. The point is that these mundane building products required high levels of quality production in standardisation, transportation and laying to ensure consistent quality performance, as testified by the structures built of sun-baked and burnt bricks dating back thousands of years.

The *Tang Lu Shu Yi Za Lu Men* (Introduction to the Laws of the Tang Dynasty: Miscellaneous Categories) set out practices in standardisation and calibration, compiled between CE 635–640, stating that measuring tools should be checked annually in August with seals fixed to demonstrate compliance. If the seals were broken, then the tools should not be used. This standardisation extended to products themselves, including the Terracotta Army being assembled from standardised parts.

In the ninth century, across the Islamic Empire of the Caliphate, covering the cities of Baghdad, Basra, Damascus and Cairo, guilds known as *naqabat*, started to appear. Carefully policed at a local level to ensure fair prices and quality, the *naqabat* each had its master (*mu'allim*), journeymen (*sani'*), and apprentices (*mabtadi*). Such *naqabat* were slowly overtaken in the nineteenth century as European industries imposed their own standards and processes.

Likewise, guilds appeared in India in the seventh century, Japan (known as *Kumai*) in the fourteenth century and China (sometimes referred to as *huiguan*) in the Qing dynasty of the sixteenth century. In Europe, guilds emerged between the twelfth and the nineteenth centuries.[6]

Guilds across the world had a wider set of objectives and often had charitable goals to teach and even care for craftsmen and their families within their communities. The specific area of guild quality standards is likely to have impacted upon building since craftspeople were engaged as contractors on specific parts of building construction, such as decorating interiors. Their quality control is likely to have influenced builders.

Abu al-Farghānī, also known as Alfraganus in the West, was ostensibly a ninth-century astronomer at the court of Abbasid in Baghdad, whose calculations of the Earth's circumference were used 600 years later by Columbus as the basis for calculations in his nautical explorations. Alfraganus was also the construction manager for the Nilometer building that was built around CE 861 in Cairo on the southern tip of Roda Island. Descending stone steps underground, a modern visitor can still see the central, engraved stone, octagonal column that measured the rise and fall of the river, which was an essential scientific device for foretelling the next season's impact of flooding or drought on the agricultural plains and the harvest. Water entered the underground chamber through three tunnels and the column was divided into nineteen cubits with the 16th mark shown to be the optimum water level. The quality of engineering to design such a sophisticated and practical structure that has remained for over 1,000 years, is a demonstrable testament to Alfraganus' mathematical skills and construction quality management.

In CE 960, the Emir Adud al-Dawla commissioned the building of a dyke that still stands today at the village of Band-i-Amir on the River Kur, twenty miles north-east of Shiraz in Iran. The Emir was noted for being a great patron of public works, ranging from hospitals to caravanserai (roadside inns) that as the nineteenth-century traveller, Baron C.A. De Bode noted:

> Bend-Amir consists of sixty houses, with twenty-one water mills erected on the river of the same name (the Kur). Here is the famous dyke which was constructed in the tenth century by Amir Uzun-Deylemi, from whom the river Kum Feruz, after its junction with the Murgab (the Polvar and Medus of the ancients) has derived its name, Bend-Amir signifying the Dyke of the Chief. A flat bridge of thirteen arches is thrown over the stream, the waters of which form a beautiful cascade just under it.[7]

The *Yingzao Fashi* (State Building Standards) is the oldest technical building manual from ancient China, written in 1100 by Li Jie, who became the Director of the Palace Building. He set out in astonishing detail, the design and construction principles in his various technical drawings. Firing bricks and tiles by a combustion of 'firstly grass then wormwood, later pine branches'[8] produced dense smoke that impregnated carbon particles onto the brick and tile surface, lowering porosity and delivering high quality bricks and tiles.

At Sacsayhuamán, near Cusco in Peru, is a citadel, started in 1100 and expanded by the Incas in the thirteenth century. Its impressively huge walls are built from apparently randomly shaped stones that fit flawlessly together (Figure 3.2). Some stones weigh several tonnes and most have multiple sides and yet were somehow hauled from quarries and finished to the highest precision before being placed next to the neighbouring stones so perfectly that a piece of paper cannot be inserted between them. Different theories have been offered and the most likely seems that the walls were created using a wooden template of the shape of the next rock required in the sequence. This would be used to inscribe

Figure 3.2 Sacsayhuamán walls in Peru, started in 1100.
Source: pixabay.com.

the outline on a rough cut rock and then methodically worked by pounding the edges to slowly meet the inscribed lines. It would have taken phenomenal amounts of hours, strength and skill by many artisans, with one slip causing a rock to be rejected for a particular place. However, this might then have been reused in a different place later. To put this into context, the total weight of stones at Stonehenge may be around 680 tonnes,[9] at Sacsayhuamán, the weight of stone in three walls is 16,626 tonnes.

The 'Temple City', Angkor Wat in Cambodia (Figure 3.3), was started in the late twelfth century and it is thought a twenty-two-mile canal was built to ferry the 5–10 million sandstone blocks for its construction.[10] The Khmer Empire under King Suryavarman II ordered the building of the temple on the 163 hectare site in the capital city of Angkor. The central site contains a temple constructed of sandstone blocks weighing up to 1.5 tonnes and made as smooth as polished marble, and there is little evidence of any mortar between the stone blocks. The skills and knowledge of stonemasonry date back thousands of years. It takes a considerable amount of experience to identify the right stone from the grain and then use a line of chiselled holes that when split to create a cleavage plane, from which the surfaces can then be further worked to create the building blocks.

Figure 3.3 Angkor Wat in Cambodia, started in the late twelfth century.

Source: pixabay.com.

The great Gothic cathedrals and churches of France started with St Denis Abbey in the north of Paris, after Abbot Suger began a novel renovation. The wooden basilica was too small to fit the increasing congregation who celebrated the regular festivals and feasts. The Abbey was also becoming more famous as it bore the name of the patron saint of France, Saint Denis, after he had been ordained by King Louis VI. The Abbey was generally believed at that time to have been consecrated by Jesus himself, so Abbot Suger could not simply demolish the building and start again. In 1135, the Western façade was rebuilt, which took five years under the design of two master masons. Despite some changes in the design, encompassing the holy Trinity through three doors, three vertical strata and a number of triple arches, the approach was typically Romanesque.

Suger was determined that the interior should be bathed in light through stained glass but building the thick walls in a Romanesque style would prevent a large stained glass window in the eastern chevet of the church. While narrow columns could support the roof with pointed arches instead of rounded ones, flying buttresses would also be needed on the outside to ensure stability for the walls.[11] These design features had been used elsewhere but St Denis was the first church (it was not created a cathedral until 1966) to use them together to achieve the tall, light-filled interior we have come to see in many cathedrals.

Abbot Suger was called away as Regent of France, in the absence of Louis VII, who had gone on the Second Crusade, so the church renovation nearly came to a grinding halt. With the Abbot's death in 1151, it was left until 1281 to complete the nave, but the initial renovation techniques became the new 'Gothic' style across France and later most of Europe. There followed Noyons in 1140, then Laon (1145), St Malo and Dol-de-Bretagne – both in Normandy (1155), Soissons (1158), Rouen (built in 1160, see Figure 3.4), Notre Dame in Paris (1163), and Chartres (1175). Within 35 years, eight new-style Gothic cathedrals had begun to be built.

The number of cathedrals helped to develop the quality control role in France of *Maître d'œuvre* (master of works and similar role to the British clerk of works). In 1145, the Count Guillaume d'Alençon founded Perseigne Abbey, as a Cistercian abbey, and Bernard of Clairvaux sent him twelve monks, two novices and twenty-one converses or lay monks, under the direction of Abbot Erard to help with the construction. These monks, some of whom were *Maîtres d'œuvre*, were renowned for their expertise in project management, architecture and construction quality.

Around the same time that Angkor Wat was being built in Asia, in 1179, the Naviglio Grande canal in Italy was started to connect the River Ticino near Oleggio, through to Milan. It took thirty years to complete and the canal served as an irrigation conduit but later transported goods through 150 kilometres of interconnected canals to Switzerland. The Naviglio Grande and its counterparts were designed to provide a complete canal system around Milan.

In Ethiopia, eleven rock-carved churches were commissioned by King Lalibela in the twelfth and thirteenth centuries, with Bet Medhane Alem (Saviour

Figure 3.4 Rouen Cathedral in France, with renovation started in 1135.

Source: pixabay.com.

of the World) believed to be the largest rock-hewn church in the world, measuring 33.5 metres by 23.5 metres and 11.5 metres high. From the natural rock,
the builders cut away the red volcanic tuff to shape a church resembling more
like a Greek temple with thirty-four rectangular columns around its walls. It
remains attached to the natural rock only at its base.

The earliest continuously inhabited house in England is claimed to be Fyfield
Hall[12] in Essex, dating to 1167–1185 with one timber post dating to the ninth
century. However, the Grade I-listed Luddesdown Court, near Cobham, Kent.
also claims the title, based upon it being held by William the Conqueror's half-
brother, Odo, until he was disgraced in 1082. Timber-framed buildings typically
used a pre-fabricated approach. Cross and wall frames in oak were built flat on
the ground, held together with mortice and tenon joints, secured with oak pegs
hammered through them. Occasionally, lap joints were used around beams and
between collars and rafters. Once erected, the panel spaces in the frames were
then infilled using wattle and daub (flexible hazel or oak sticks interwoven and
covered in a sticky mixture of clay, animal dung and straw) or bricks. If the
frames were left uncovered, then we still can see them today in their natural
state, although many dating to Tudor times have been painted in black and
white. Examples of these medieval buildings can be seen across many towns and
cities in England, including Durham, York, Chester, Shrewsbury, Canterbury
and Faversham.

The Qutub Minar[13] in Delhi, India, is the tallest minaret in the world made
of bricks. In 1199, Qutb-ud-din laid the foundations of a majestic masonry tower
which is 72.5 metres in height with subsequent storeys added over the next 200
years. The diameter at its base is 14.32 metres tapering to just 2.75 metres at the
top with a 63 cm tilt towards the south-west. Its design was based upon Afghani-
stan's Minaret of Jam, and the minaret and surrounding buildings demonstrate
the construction techniques of indigenous traditions and the development of
Islamic architecture in India. An inner shaft provides a staircase to the top and
is connected using timber beams to an external shell with exquisite carvings on
different stone façades. The lower three storeys are made from sandstone and
the top two storeys from amalgamations of sandstone and marble.

York Minster[14] in England (Figure 3.5) can trace churches built on its site
back to CE 633 but the current building was started c.1225 and completed in
1472. Archbishop Walter de Grey started the rebuilding of the South Transept
around 1220, on top of the existing church, but whoever was the architect built
it some three feet off-centre as anyone looking at the Rose window can see. Ten
years later, the North Transept was started and it is likely just one master mason
was in charge of the project until its completion in 1260. While also being
slightly off-centre, it is more carefully hidden. The first named mason was
'Simon le Mason', who designed and built the nave between 1291 and 1360,
with the north wall around one foot lower than the south wall, probably due to
an inability to checks levels from the old Norman nave between them. The
choir was also poorly set out by William de Hoton with the tower unsighted
from the old Norman choir that was being replaced. By the time that the new

Figure 3.5 York Minster, England, showing its flying buttresses. A church has been on this spot since CE 627 and the present structure dates from *c.*1225.

choir connected with the tower, it was out by three feet. While the overall Minister is magnificent, the lack of one controlling mind in design and the apparent basic surveying mistakes have resulted in quality management mistakes for those with sharp eyesight.

Master masons acting as architects, structural engineers, construction managers and quantity surveyors were also the quality control inspectors, paid on the amount of approved work that was completed to the required standard. They acted as a contractual backstop against poor quality but no doubt compromises had to be made and sometimes defects were not even seen until decades after completion. Mason marks were made on quarry stones to identify their source, assembly marks to ensure correct construction of adjacent stones and bankers' marks, which were the individual stonemason's marks to identify them personally. The 'Banker' name came from the stone bed or bench that the mason used to work the stone. Day-to-day accounts for a construction project were usually managed by the master mason who would sign the monthly accounts with his name and mark. Usually a temporary wooden building was erected on the building site, called a 'lodge', for the masons to store materials, use as a workshop and socially enjoy companionship. When Abbot Suger at Saint Denis wanted more light to enter the Abbey, it would have been the master mason who would have been tasked to design and build it, in a structurally sound way.

King Wenceslaus the First issued a proclamation on mining for Jihlava (now in the Czech Republic), *Jus civiim et montanorum* in 1249, which included detailed inspection responsibilities of a land registrar for mining. Additional laws were added and, in 1305, the *Constitutiones iuris metallica* created a comprehensive mining system that would be the basis of regulations for more than three centuries across central parts of Europe. Land registrars appointed 'top climbers' who would inspect each mine underground, adjudicating any disputes and supervise the performance of the extraction of precious metals.

While many impressive bridges have been built throughout history, the bridge over the River Adda at Trezzo in Italy, built between 1370 and 1377, had a span of a single arch that was 236 feet long with a rise of just seventy feet. That span ratio of 0.3 is incredibly low by historical standards, with most Roman arches semi-circular in ratio, i.e. 1:1. The bridge at Avignon had a ratio of 0.83. The Trezzo sull'Adda bridge held the record for having the largest single span for over 400 years until the metal Wearmouth Bridge of the same span was erected at Sunderland in England, in 1796.

In England, in Richard II's time, the *Regius* text,[15] also known as the *Halliwell Manuscript*, so called as it was rediscovered by James Halliwell in 1840, was probably written by a West of England clergyman and sets out in poetical form the rules for stonemasons engaged with building cathedrals and churches. Written around 1390 in Middle English, it states the role of the master mason and their solemn duty to honestly supervise the workers and ensure that they are fairly paid:

> The first article of this geometry;
> The master mason must be full securely
> Both steadfast, trusty and true,
> It shall him never then rue;

And pay thy fellows after the cost,
As victuals goeth then, well thou woste; [knowest]
And pay them truly, upon thy fay, [faith]
What they deserven may; [may deserve]
And to their hire take no more.

Running to some 974 lines of script on sixty-four vellum pages, it is one of the earliest 'charters' for codifying rules for builders.

It was in the same year, 1390, that Geoffrey Chaucer, English poet and philosopher, was appointed Clerk of King's Works, overseeing the erection scaffolds for a tournament. The mandate dated 1 July 1390, to the Exchequer to allow Chaucer the costs of building scaffolds in Smithfield for the jousts of the preceding May, stated:

> For Geoffrey Chaucer. Richard by the grace of God king, etc., to the treasurer and barons of our exchequer, greetings.
>
> We command you that in the account which our dear Squire Geoffrey Chaucer, Clerk of our Works, is to present you in the execution of his office you will allow him on his oath the costs made for the scaffolds in Smithfield, which he had made for us and for our very dear companion the queen at Smithfield in the Month of May now passed.
>
> Given under our privy seal at Westminster, the first day of July in the fourteenth year of our reign.[16]

This role, Clerk of King's Works, was an authoritative figure on a construction project representing the client who was paying and ensuring that work was carried out to the necessary standards and regulations. Richard II also appointed Chaucer as Clerk of King's Works for the Palace of Westminster, the Tower of London and a number of other royal residences, being paid a daily wage of two shillings (£36 10s. annually, when a labourer would earn £2 per year and a stone mason £8 per year[17]). A year later, he was appointed to manage the repairs to the King's chapel of St George at Windsor Castle.

The Forbidden City in Beijing was constructed between 1406 to 1420 and consists of 980 buildings covering some 180 acres and supervised by the Yongle Emperor's Ministry of Work of the Ming dynasty. The meticulously designed and executed construction needed 100,000 skilled artisans and around one million labourers and was carefully planned to conform to the 2,500-year-old *I Ching* or Book of Changes that sought to harmonise humans and nature under the traditional Chinese Confucian culture. As the Jesuit missionary, Father Pierre-Martial Cibot wrote in 1785, upon visiting the Forbidden City:

> The palaces of the emperor are real palaces and bear witness to the grandeur of the lord who inhabits them by the immensity, symmetry, elevation, regularity, splendor and magnificence of the innumerable edifices which compose them. The Louvre would largely stand in one of the courtyards of

the palace of Peking, and there are many from the first entrance to the more secret apartment of the Emperor, not to mention the lateral edifices. All the missionaries we saw coming from Europe were struck by the air of grandeur, wealth and power of the palace of Peking. All have confessed that if the various parts which compose it do not enchant sight, as the finest examples of great European architecture, their whole constitutes a spectacle to which nothing of what they had seen before had prepared them.[18]

The royal Inca estate of Machu Picchu in Peru (Figure 3.6), was built around 1450 and 1460 and it has been estimated that around 60 per cent of its construction is unseen below ground, with deep foundations.[19] The Incas used only bronze tools and hard stones to shape the granite, porphyry, limestone and basalt and used the technique of dry ashlar to piece together tightly knit walls without the aid of mortar (Figure 3.7). Using these centuries-old dry ashlar techniques also used at Sacsayhuamán, these free-standing walls are more resistant to the local earthquakes, with the stones in the walls known to slightly jump from the vibrations, but remain undamaged.

The architecture of the city of some 200 structures flows around the rocky landscape, 2,420 metres above sea level, using terracing. Typically the houses had sixty degree thatched roofs to throw off the regular down pour of water and

Figure 3.6 Machu Picchu in Peru, started in c.1450
Source: pexels.com.

Figure 3.7 Immaculately cut stones fitted together at Machu Picchu.

Source: pixabay.com.

an underground drainage system carved out of the rock. Temples and city walls were built from the igneous rock, which was shaped into immaculately fitting stones of rectangular rows and polished using sand to a fine finish. Most of the rock was local to the Andes mountains but some was brought by llama or dragged by people up the mountainside.

The impressive standard of building in such a dramatic location with the simplest of tools and technology is astonishing. Given the speed of building the city, it took talented, skilled craftspeople and builders to create a built community perched on the top of the mountain, and it needed architects intimately involved daily with the construction, in order to achieve the incredible quality control results. It is a testament to the skills that were developed using trial and error and then passed on orally through generations, which were improved over decades of training and practice.

In 1538, the role of Ordnance Surveyor[20] was created in England, appointed by the Crown, acting as a chief engineer, with duties of quality control for military munitions and infrastructure, such as forts and barracks to support the Army and the Royal Navy. The role reported through to the Board of Ordnance with its headquarters in the Tower of London.

The gleaming white marble, Taj Mahal,[21] in India, built between 1631 and 1648 by Mughal Emperor Shah Jahan, was dedicated to Jahan's favourite wife, Mumtaz Mahal. It took 22 years to complete by 22,000 labourers using 1,000 elephants, and was constructed of translucent white marble, brought to Agra from the Indian region of Rajasthan.[22] The marble was then inlaid with semi-precious stones, including jasper, lazuli, sapphire, turquoise, jade, carnelian and lapis to produce a stunning impact throughout the day as the reflected light moves through an arc of pink in the morning, white at mid-day and gold in the evening sunset. However, modern industrial air pollution has caused havoc with the white stone.

The humble, five-storey building known as the Flaxmill Maltings in Ditherington, near Shrewsbury in England, was built between 1796 and 1797 but has enormous historical importance. The steam-powered mill was originally built to spin flax fibre to make linen and later was converted into a factory to make malt for the brewing industry. It is the world's first iron-framed building and, as such, the antecedent of all skyscrapers. Charles Bage, the architect, had been inspired by the designs by William Strutt, who had developed iron frames, as opposed to timber frames, in buildings to improve their fire resistance. Cast iron beams are supported by cast iron columns, tied together by wrought iron tie rods that run axially between the beams that support the brick arch ceilings and floors. With virtually no timber exposed internally, it was a robustly fire-resistant design. The innovative use of the cruciform cross-section columns made quality control better, as defects could be identified more easily.

The lesson for today's quality professional is that they need to appreciate that new technologies in design, like those used in Hagia Sophia, St Denis Abbey, Bet Medhane Alem and the Flaxmill, mean they need to adapt quality management professional techniques to assess quality risks and check that they have been suitably mitigated.

Notes

1 *Panditabhushana V-Subrahmanya Sastri*, B. *Brihat Samhita of Varaha Mihira* LVI.31, LVII 1–7. (Trans. 1946). Retrieved from https://archive.org/stream/Brihatsamhita/brihatsamhita_djvu.txt
2 Nene, A.S., 'Rock engineering in ancient India' (2011). Retrieved from https://gndec.ac.in/~igs/ldh/conf/2011/articles/Theme%20-%20P%202.pdf
3 Needham, J., *The Shorter Science and Civilisation in China* (Cambridge: Cambridge University Press, 1994), pp. 145–147.
4 The Holy Quran, Chapter (61) sūrat l-ṣaf (The Row), Verse 61:4, Retrieved from http://corpus.quran.com/translation.jsp?chapter=61&verse=4
5 Ibid., Chapter (28) sūrat l-qaṣaṣ (The Stories), Verse 28:38. Retrieved from http://corpus.quran.com/translation.jsp?chapter=28&verse=38
6 *Encyclopedia of the Social Sciences*, 'Guilds' (New York, 1938), vol. VII, pp. 204–224. Retrieved from https://archive.org/details/encyclopaediaoft030467mbp/page/n3
7 Baron de Bode, C.A., *Travels in Luristan and Arabistan* (1845), vol. 1, p. 171. Retrieved from https://books.google.co.uk/books?id=i_gqUpmQRIwC&pg=PA97&source=gbs_toc_r&cad=4#v=onepage&q&f=false
8 Guo, Q., *Tile and Brick Making in China: A Study of the Yingzao Fashi* (2000). Retrieved from www.arct.cam.ac.uk/Downloads/chs/final-chs-vol.16/chs-vol.16-pp.3-to-11.pdf
9 Daw, T., 'How many stones are there at Stonehenge?' (2 March 2013). Retrieved from www.sarsen.org/2013/03/how-many-stones-are-there-at-stonehenge.html
10 Ghose, T., 'Mystery of Angkor Wat Temple's huge stones solved'. *Livescience*. Retrieved from www.livescience.com/24440-angkor-wat-canals.html (accessed 31 October 2012).
11 Calkins, R.G., *Medieval Architecture in Western Europe: From* A.D. *300 to 1500* (New York: Oxford University Press, 1998), pp. 172–173.
12 Fyfield Hall. Retrieved from www.fyfieldhall.co.uk/history
13 Wikipedia. 'Plaque at Qutub Minar'. Retrieved from https://en.wikipedia.org/wiki/File:Plaque_at_Qutub_Minar.jpg
14 York Minster. 'A brief history of York Minster' (2018). Retrieved from https://york-minster.org/discover/timeline/
15 Halliwell, J., reproduced by Pietre-Stones (1840). Retrieved from www.freemasons-freemasonry.com/regius.html
16 Crow, M. and Olson, C.C., *Chaucer Life-records* (Oxford: Oxford University Press, 1966).
17 Medieval prices and wages. Retrieved from – https://thehistoryofengland.co.uk/resource/medieval-prices-and-wages/
18 *Mémoires concernant l'histoire, les sciences, les arts, les mœurs, les usages des Chinois* (Peking, 1782), vol. XIII. Retrieved from https://gallica.bnf.fr/ark:/12148/bpt6k114468v/f8.image
19 Adams, M., 'Top 10 Machu Picchu secrets'. *National Geographic*, November 2018. Retrieved from www.nationalgeographic.com/travel/top-10/peru/machu-picchu/secrets/ (accessed 31 July 2000).
20 Sainty, J.C., *Ordnance Surveyor 1538 to 1854* (London: Institute of Historical Research, 2002). Retrieved from www.history.ac.uk/publications/office/ordnance-surveyor
21 UNESCO. *Taj Mahal*. Retrieved from https://whc.unesco.org/en/list/252
22 Visit India. *History of Taj Mahal*. Retrieved from www.visittnt.com/taj-mahal-tours/history-of-taj-mahal.html

4 Modern construction, 1800–2000
Quality and lean construction

From the start of the Industrial Revolution towards the end of the eighteenth century, the construction industry faced increasingly greater challenges as cities expanded and the demand grew for new living and working spaces. The nineteenth century brought a growing awareness of material quality and returned to a theme throughout history of craftsmanship in buildings.

In Washington, DC, the Executive Mansion or President's House or President's Palace, as it was variously called,[1] originally was completed in 1800. The construction of the 'White House' (Figure 4.1) began on 13 October 1792 with

Figure 4.1 First known photograph of the White House in 1846 taken by John Plumbe.
Source: Library of Congress/John Plumbe.

George Washington laying the first cornerstone. Aquia sandstone was quarried from Government Island, located in Aquia Creek that runs into the Potomac River, after being purchased by the French engineer, Pierre Charles L'Enfant on the approval of George Washington. Irish-born architect James Hoban designed the building in a neoclassical style inspired by the Roman architect, Vitruvius and the Renaissance-era architect, Andrea Palladio. He had intended for it to have three storeys with nine bays, but this was altered to two storeys with eleven bays.

Slaves were used to quarry the stone and, in 1794, seven stonemasons travelled from Edinburgh in Scotland. Oxen pulled the large stones to the edge of the river to be loaded onto barges and paddled forty miles to the city of Washington. Master mason Colin Williamson oversaw the stones being laid out and numbered before being assembled in precise order and whitewashed to protect the porous stone from winter freezing damage.[2] This technique was designed to weather the building but keep the stones' crevices infilled, however, it was decided to regularly whitewash the stones' surface to keep it fresh and the nickname of the 'White House' became official in 1901. Whitewashing stone goes back to *c*.3517–3358 BCE in Mesopotamia and may have its origins from antibacterial properties in the whitewash as well as aesthetic reasons. The White Temple at Uruk (modern-day Warka), dedicated to the God Anu, was whitewashed, so it could be seen from miles away.[3]

The Code Napoléon, published in 1804 in France, swept away the old feudal order and included strict responsibilities for owners and their agents, including architects and the builders: 'If the edifice, built at a set price, perish in whole or in part by defect in its construction, even by defect in the foundation, the architect and the contractor are responsible therefore for ten years.'[4] Such responsibilities were extended with amendments to the Code over the following years.

The first tunnel under a river was the Thames Tunnel in London, started in 1825. Until that time tunnels had been built for thousands of years but usually with a 'cut and cover' method and not under a river.[5] Trade in London was a crowded affair with up to 3,000 tall ships on the river each day and to bring cargo across the river was difficult. Marc Isambard Brunel (Isambard Kingdom Brunel's father) designed an innovative boring shield; thirty-eight feet wide by twenty-two feet high and it was to become the prototype for all future tunnelling shields. The idea came to Brunel when he noticed how ship worms ate their way through timbers with their head protected by a hard shell. Miners worked in cages in three- to four-hour shifts, digging through the rock in four inch strips. Behind them, workers would lay bricks to maintain a robust construction. However, the dangers the men faced were awful. Raw sewage would spray into their faces and methane gas would ignite into flames from miners' lamps. The 1,200 feet tunnel flooded five times, with Isambard Kingdom, who was working as an engineer in the tunnel, only just surviving with his life on 12 January 1828. The tunnelling was suspended for seven years when the project ran into financial troubles, only re-starting when Brunel secured a loan of £247,000 from the Treasury.

It took until 1843 before the tunnel opened to acclamation and 50,000 people paid a penny to walk through it. After three months, one million had visited; equivalent to half the population of London at that time. Designed as a pedestrian and horse and carriage tunnel, it was added to the underground train network in 1869.

In his short book, *Hints to Persons about Building in the Country*, in 1847, the young US author Andrew Jackson Downing,[6] the founder of American landscape architecture and a great advocate of the Gothic Revival in the United States, made frequent exhortations for building quality to his readers and budding builders. In the Introduction, the failure to understand the practical considerations of house building would totally undermine the decorative beauty and customer satisfaction:

> People whose tempers are disturbed by leaks and offensive smells, by damps and smoky houses, or even by partial failures in design, will become proportionally blind to the numerous merits which may still remain; and the architect at the moment of signing his last certificate 'that the contractor has fulfilled all his duties in the most complete and workmanlike manner' may be in effect signing the declaration of his own inefficiency and unconsciously entering on a period of much trouble....

Roofs made of tin or zinc must 'specify the quality' and floors were 'to be laid of the best quality' with further advice on how to construct stairs of 'better quality' and selecting the 'quality of bricks' and even the 'quality of the soil immediately about site of the house will not be overlooked',[7] as linking houses to their surrounding landscape was an important part of his philosophy.

In 1851, the Crystal Palace, designed by Joseph Paxton, was erected in Hyde Park in London (later moved to Sydenham Hill in 1852) for the first international Great Exhibition. The sophisticated 'greenhouse' was constructed from two parallel rows of 500 iron columns, each supporting 2,224 trellis girders covered with 300,000 sheets of glass with a fixed ridge and furrow roof that encased existing mature trees in the park.

Ventilation was by a novel design of manually operated horizontal metal louvers opened and closed through the turning of a wheel and cord mechanism designed by Fox, Henderson & Co., the contractors. This mechanism operated twenty-four ventilator units from one position.

The glass was specially made at 1.2 metres (four feet) long and just 2 mm (1/13 in) thick, which conformed to Paxton's four feet modular design so that the structure could be built as if in an assembly line. The lightweight design meant that Fox, Henderson & Co. constructed it using mainly manpower and horsepower, without the need for machines, in eight months. Wheeled trolleys ran on the gutters, using them like rails, which eliminated scaffolding for the glaziers. Some eighty men could fix 18,000 panes of glass in a week and this efficiency transcended to the site hoarding being re-used as the building's flooring. It was an iconic model for standardisation. It caused a huge debate between architects and engineers and while the public loved it, it was castigated by some in the architectural profession. It was formally opened on 1 May 1851 by Queen Victoria, with six million visitors in the first five months viewing 13,000 exhibits from around the globe, including the world's largest diamond, the Koh-i-Noor.

In 1853, François Coignet was the first to use iron in concrete, *béton armé*, to build a four-storey house at 72 rue Charles Michels in Paris. Tragically, the building has been allowed to fall into ruin.

In 1854, Sir William Fairbairn, a Scottish civil engineer and the third president of the Institution of Mechanical Engineers, wrote about the superiority of wrought iron over cast iron and warned about hidden quality defects in cast iron with 'the impossibility of discovering imperfections that may lie concealed under the surface of a casting and which frequently baffle the scrutiny of the keenest observer'.[8]

While artificial stone has been around since Georgian times in the UK, a British patented, concrete paving stone won an 1862 CE International Exhibition Prize Medal in London and boasted that it had all the enduring properties of the 'Old Roman Concretes or Mortars'. Comprising sand, chalk or other mineral substance, it was mixed with a siliceous cementing material, pressed into blocks or moulds and immersed in a solution of chloride of calcium. It was claimed the artificial stone was cheaper than quarrying real stone.[9]

Peter Ellis was an architect in Liverpool and noted that the typical office buildings being built had small Gothic windows that allowed little light inside for workers. In 1864, he designed the Oriel Chambers[10] for the Reverend Thomas Anderson, with the first-ever glass curtain wall and an iron frame and two years later, used a similar design for 16 Cook Street, both in Liverpool. It was scorned at the time by the architectural establishment but helped to inspire skyscraper architects in Chicago.

In 1869, Washington Roebling was appointed Chief Engineer of the proposed Brooklyn Bridge in New York. He had inherited the design from his father, John, but only a year later suffered a debilitating and paralysing injury from decompression sickness and his wife, Emily, took over the day-to-day project management and quality control of the construction until its inaugural opening by President Chester Arthur in 1883. With Washington unable to attend the ceremony, Emily accompanied the President as the first to formally cross the bridge. Emily Roebling is recognised as being the Acting Chief Engineer for the longest span suspension bridge (at the time) and the first steel-wire suspension bridge in the world.[11]

Between 1871 and 1901, the Philadelphia City Hall was briefly the world's tallest building before being surpassed by the Eiffel Tower. Constructed with eighty-eight million bricks,[12] its walls are up to 6.7 metres thick to support the 167 metres height.

A US patent was filed in 1873 by Alfred Hall from New Jersey on how to improve the 'coloring processes of bricks' but it was contingent on the 'quality of the brick' so 'as to stand the heat required to flux the mineral coloring matter without the use of a lead or other flux which will produce glazing'. Another US patent in 1886, by Henry Dickson of Pennsylvania, proposed that pulverising slate waste and mixing with aluminous clay before tempering and placing in a kiln, produced a 'very superior brick, tile or drain pipe'.

In the UK, the Clerk of Works Association started in 1882 (later the Incorporated Clerk of Works Association of Great Britain in 1903 and changed to the Institute of Clerks of Works of Great Britain Incorporated in 1947 and finally the Institute of Clerks of Works and Construction Inspectorate of Great Britain Incorporated in 2009), as the collective professional association of Clerks of Works, who represented the client rather than the contractor. These early professional inspectors carried out assessment of construction works and held a high level of expertise to ensure that standards and specifications were being adhered to, as well as sometimes project managing work.

In 1885, the ten-storey Home Insurance Building in Chicago (Figure 4.2) was constructed (and unfortunately demolished in 1931) and seen as the father of modern-day skyscrapers that dominate the skylines of many cities around the world. Engineer architect, William LeBaron Jenney, who was educated at the same establishment as Gustave Eiffel, at the École Centrale, Paris, thought that an iron skeleton could support a tall building, so legend proclaims, after seeing his wife place a heavy book on top of their birdcage.

The Eiffel Tower (Figure 4.3) was constructed between 1887–1889 to celebrate the centenary of the French Revolution.[13] Twice as high as the Great Pyramid of Giza with a total elevation of 324 metres, four lattice girders pillars taper and merge to form a vertical tower of 18,038 pieces of wrought iron, held together by 2.5 million rivets with holes bored to 0.1 mm tolerance. Foundations were set below the water level that reached fifteen metres down and the first staging required the slant of the girders to be erected using hydraulic jacks and boxes of sand to establish the correct angle, with the pillars correctly positioned to one millimetre accuracy. Cranes were used to construct the second stage, along the route of where elevators would eventually be fixed, before the tower was erected on top.

All the iron elements were constructed at Gustav Eiffel's factory at Levallois-Perret on the outskirts of Paris. The contractors were experienced in building metal viaducts and the iron components were held together in the factory by nuts and bolts before each one was replaced with a rivet on site in record time. It took just two years, two months and five days to complete construction. Some 300 well-known Parisian artists and intellectuals had written a letter of protest against the design in February 1887 calling it 'useless and monstrous' but the tower grew to be an iconic image of the city.

In the eighteenth century, an iron smelting works was set up by Korobov in the south Urals near Lake Kasli and was sold on to Demidov, one of the largest metallurgy businessmen, who developed the smelting works for producing cast iron cannons and cannonballs. By the late eighteenth century, architectural pieces were being produced to the highest standards, so much so that the Kasli iron pavilion,[14] cast in 1898–1899 from the design by the Russian architect,

Figure 4.2 The 'father' of modern skyscrapers; the 1885 Home Insurance Building in Chicago.

Source: New York World-Telegram and the Sun Newspaper Photograph Collection (Library of Congress).

Figure 4.3 The iconic Eiffel Tower assembled from 18,000 wrought iron pieces and 2.5 million rivets.

Yevgeny Baumgarten, was showcased for the 1900 International Exhibition in Paris. The intricate designs of motifs of ancient Russia and Byzantium won the Grand Prix Crystal Globe and the Big Gold Medal.

The twentieth century saw a massive expansion of construction with the technological material developments in reinforced concrete, glass, steel and aluminium and radical, improved processes for more efficient off-site fabrication and assembly.

The Robie House in Chicago, Illinois, built in 1910, was one of Frank Lloyd Wright's 'prairie houses', showered with praise for its open-plan, light-filled and expansive interior. Built by contractors H.B. Barnard Company in Roman brick, the vertical joints were coloured the same as the bricks, while the horizontal mortar was traditional cream-coloured. It was a clever trick on the eye to extenuate the horizontal lines of the design. It was considered to have established a new form of house design in the USA.

The Fallingwater residence in Pennsylvania that Wright designed in 1935 and influenced by Japanese design principles of harmony with nature, cantilevered over a waterfall with layers of cream-coloured, reinforced concrete floors. These cantilevered floors, however, were a source of contention with the family's own consulting engineers, who quietly added additional reinforcement to the cantilever design without telling Wright. However, even this strengthening was not sufficient, with modern-day engineers finding that it was close to its failure limits, and required additional beams. The difference of opinion with engineers did not stop Wright from going on to become one of America's greatest architects, designing New York's Solomon R. Guggenheim Museum in 1959 with its distinctive spiralling white ribbon of concrete façade becoming progressive wider as it rose upwards.

The Waldorf Astoria Hotel was disassembled to make way for the new Empire State Building (ESB) with the ground-breaking of the site beginning on 17 March 1930,[15] constructed at the astonishing rate of approximately one floor per day and was structurally finished on 11 April 1931; twelve days ahead of schedule and under budget, with a final gold rivet fired into place.

The innovative, fast-track construction process meant that the construction started before the design was complete with the general contractors, Starrett Brothers & Eken, taking a new approach of having all key construction equipment, custom-made rather than renting existing equipment. The logistical programming was a *tour de force* with 60,000 tonnes of steel manufactured in Pittsburgh and just-in-time transportation to New York, so that girders could be riveted in place, eighty hours after coming out of the furnace.

Instead of storing bricks in the street below and using wheelbarrows to move around the site, they were delivered into the basement by chute, with two hoppers feeding bricks into the railway cars that moved them in a horizontal direction to hoists for transporting vertically to the required floors. The ESB contained 62,000 cubic yards of concrete and 200,000 cubic feet of Indiana limestone and granite. The water system was built inside of the main structure instead of on top of the roof, as was the norm in those days. Originally conceived at eighty storeys, in the end it topped out at 102 storeys, 448 metres tall, as the design was repeatedly adapted in competition with the newly constructed Chrysler Building.

There are thousands of Modernist buildings around the world that have been built in the past century that have inspired both adoration and contempt in equal measure. Most have used materials that are harsher on the eye: reinforced concrete, glass and steel.

The Villa Savoye is a Modernist take on the classic French villa, with an elegant box lifted off the ground by slender columns, built in 1931, with a roof garden. Designed by the Swiss architect and urban planner, Le Corbusier (born Charles Edouard Jeanneret), some of his famous buildings in concrete included the Cité Radieuse in Marseille, the Palace of Assembly in Chandigarh, India, in 1951, as joint-architect (with Oscar Niemeyer) of the United Nations Head-quarters in New York (Figure 4.4), completed in 1952, the Notre Dame du Haut in Ronchamp, in 1954, the National Museum of Western Art, Tokyo, in 1959, and the multi-coloured Pavillon Le Corbusier, Zurich, in 1967, completed two years after his death and conceived of as a museum dedicated to him.

The Seven Great Wonders of the ancient world were:

- the Great Pyramid of Giza;
- the Hanging Gardens of Babylon;
- the Temple of Artemis;
- the statue of Zeus at Olympia;
- the Mausoleum at Halicarnassus;
- the Colossus of Rhodes;
- the Lighthouse of Alexandria, devised by the Greek historian Diodorus Siculus (Diodorus of Sicily) in 100 BCE.

Only the Great Pyramid survives today and they were all located around the eastern side of the Mediterranean Sea, except for the Hanging Gardens (and there is some scepticism about its existence). Scroll forward 2,000 years and in 1955, the American Society of Civil Engineers in its inimitable wisdom chose the US Seven Modern Wonders of Civil Engineering, with its number one as the Chicago Sewage Disposal System. While it provides an admirable utilitarian service to the people of the great city of Chicago, there is little in its appearance to warrant inclusion from an aesthetic point of view. Yet aesthetic appearance is one quality aspect that most clients will care about in the construction of the built environment. This example demonstrates the challenge of understanding all stakeholder opinions and requirements in reaching both design and construction decisions. There can never be perfect quality but balancing the competing quality characteristics deserves time, careful consideration and a determination to seek out customer satisfaction.

The São Paulo Museum of Art extension in 1968 was a brutalist creation by the Brazilian architect, Lina Bo Bardi. Four bright red concrete pillars are wrapped around a glass box suspended from the floor to allow pedestrians to pass freely underneath.

Figure 4.4 Le Corbusier and Niemeyer designed the UN HQ in New York, built in 1952.
Source: pixabay.com.

The 110-storey Twin Towers were part of the World Trade Center complex in New York. The North Tower was started in August 1968 and topped out on 23 December 1970 (South Tower, January 1969–July 1971). The dedication ceremony waited until 1973. The architect Minoru Yamasaki chose to use an innovative structural design of placing the elevator shifts and stairwells in the centre with a perimeter of closely spaced tubular columns, which created open-plan floors for the office tenants. The tubular design delivered 40 per cent less steel than in comparable buildings. To reach the rock twenty metres below the ground level, a 'bathtub' design was constructed to keep out the Hudson river, and prefabricated components brought in by boat allowed the lightning speed of construction. They became the tallest buildings in the world at the time with Number 1 (North) World Trade Center at 417 metres and Number 2 (South) World Trade Center, 415.1 metres tall. The robust design was strong enough for the static design loads but was never designed to withstand the dynamic loads inflicted by the terrorist attacks of 2001 that brought the towers down and killed 2,996 people.

In 1971, international architects were allowed to compete in designs for public buildings in France. The Centre Georges Pompidou was designed by an Italian, Renzo Piano, with the assistance of Richard Rogers (UK) and

Gianfranco Franchini (Italy). Taking five years to complete, it houses the public information library and stretches to five acres in area and seven storeys in height. At first sight it seems it is swathed in forgotten, coloured scaffolding with a glass tube snaking up the outside. The 'inside out' building has all the internal services built on the outside fabric of the structure, providing a new template for the architecture of museums and cultural centres around the world.

The Lloyds Building in London, opened in 1986, pushed all lifts, stairwells, water pipes, ducting and other building services to the outside to make the interior a palatial open design with escalators sweeping up and down in the centre. Even today, it looks futuristic. Richard Rogers, the architect, left the 1928 original insurance building entrance at 12 Leadenhall Street, as an incongruous preservation.

Finished in 1989, the Pyramide du Louvre (Figure 4.5) is the contrasting entrance to the grand, stone-façaded Parisian museum. Designed by the Chinese-born American architect Ieoh Ming Pei, it has 673 glass panels, held in place with steel rods and cabling that used yacht technology. The twenty-two metre-high pyramid caused consternation and criticism when the design leaked with many Parisians tired of Modernist buildings that had included La Défense business district but the pyramid has since grown on the city and the eight million tourists each year. Pei wanted the glass to be as clear as possible and after help from President Mitterand, had a new manufacturing process created using Fontainebleau white sand.

Figure 4.5 The iconic Louvre glass pyramid, finished in 1985.

Source: Gerhard Bögner, pixabay.com.

After a fire destroyed part of the huge Vitra furniture production facilities, the owners decided to build a fire station. The Vitra fire station by the Iraqi architect, Zaha Hadid, completed in 1993, in Weil am Rhein, Germany, is a striking design with an overhead wing of reinforced concrete above the entrance.

The Obayashi Corporation created the 'Big Canopy'[16] system in 1994 to protect the construction of high rise buildings from the vagaries of the weather and as an attempt to create more factory conditions. Four tower crane posts topped with a temporary roof canopy sit externally to the structure being constructed and after two floors are completed, it is jacked up. Weighing 2,200 tonnes it has self-contained cranes and hoists and it is claimed to save six months of construction time for a forty-storey structure with eight out of ten typical tower blocks able to use the 'Big Canopy' system. A mixture of pre-cast and in situ concrete is used, with material management monitoring by a bar coding system linked to a database. Wind velocity was reduced by 66 per cent and heat by 10 per cent for workers. The time of one month taken to erect the Big Canopy was more than offset by time savings from virtually no weather disruption.

The Magdeburg Water Bridge, started in 1998, in the state of Saxony-Anhalt, Germany, runs over the Elbe River and connects the Elbe-Haval Canal with the Mittelland Canal. The bridge at 918 metres is the world's longest navigable aqueduct and took six years to construct, costing €501 million with 24,000 tonnes of steel and 68,000 cubic metres of concrete.

The Sendai Mediatheque library in Japan, designed by Toyo Ito, completed in 2001, has columns separating the floors, comprised of steel tubes set a different angles but acting as earthquake resistors. The double-glazed façade allows light deep inside, and at night the building is lit like a lantern.

Lean construction

With a Master of Science (Technology) degree in Industrial Engineering and Management in 1976 from the Helsinki University of Technology, Lauri Koskela has become one of the leading advocates of lean construction. In 1992, he published the paper 'Application of the new production philosophy to construction', focused on the production part of construction. He later developed his ideas that construction should start with a firm foundation of theory to underpin how and why the design was developed.

Koskela challenged the prevailing orthodoxy that construction was simply taking inputs to convert to outputs by breaking up the components into smaller parts to project manage them. This conversion model carries with it inherent difficulties. Given the number of variables and lack of clearly defined problems that need to be solved along the way of constructing something, he believed that the underlying model was flawed: 'Thus, a methodology that is more geared towards organizing discussion, debate and argument is needed. The soft systems methodology fulfils this requirement.'[17] The conversion model with standard project management is not able to deal with the reality of construction. How to manage re-work, accidents or delays between and within processes? The impact on time, cost and quality under a general heading of 'waste' is not adequately managed. A work breakdown structure (WBS) is a useful tool but usually does not take into account whether the sum of the parts will meet the customer's actual requirements. The value add in the conversion model is not based on a conceptual approach. Lean construction thoroughly interrogates and understands the purpose of the design for the customer to identify how and where to add real value and acknowledge the customer's constraints to achieve that purpose: time, cost, location and quality standards. The contractor of the construction product and/or service needs to help the customer to understand their constraints. They may want the most aesthetically pleasing bridge until the cost estimate tempers the materials they want to use.

Koskela and other collaborators, including Glenn Ballard and Greg Howell, were interested in developing a similar approach to the Toyota Production System (TPS) that had developed in the manufacture of Toyota cars between 1948 and 1975. A long-term horizon philosophy, process management to reduce waste, valuing people and partners and continual improvement through problem solving led to just-in-time, Poka Yoke, Jidoka and other 'Toyota Way' elements. Unfortunately, consultants sometimes adopted a pick-and-mix approach, using some of the individual techniques for short-term gains without understanding the full potential and necessity of a holistic approach to TPS.

The lean construction approach has developed its own tools and techniques: Building Information Modelling (BIM), 5S, value chain mapping, Design for Manufacture and Assembly (DfMA) and the Last Planner System (LPS).

The Transformational-Flow-Value goals are consolidated and balanced in priority under lean construction whereas traditionally the waste minimisation and value maximisation elements are neglected, as shown in Table 4.1.

Table 4.1 Traditional vs lean construction

Traditional construction	Lean construction
Project planning is typically undertaken sequentially before passing to the next step	Supply chain stakeholders are involved early on in project planning
Tasks are carried out as soon as possible	Tasks are carried out 'just in time'
Procurement identifies suppliers and subcontractors through market mechanisms predominantly based on price	Time in supply chain delivery is minimised and the supply chain is partnered to optimise delivery
Large inventories are typically maintained	Optimum inventories maintain supplies
The goal is transformational to convert inputs into outputs	The goal is Transformational-Flow-Value
Product design and process design considered separately	All lifecycle stages are considered in the product and process design

Tasks are set out in LPS as promises by each individual responsible for that task. Designers, site supervisors, engineers, who are the 'last planner' in the process before work starts, decide on the conditions required before making the promise of delivery. By monitoring the root causes for any failure to deliver on time, the analysis can demonstrate overall problems in processes and allow effective action to continually improve.

Looking ahead to the next tasks improves the work flow, and workforce planning, just-in-time deliveries and last planner software linked to BIM models can highlight potential errors in placing the wrong components into work packages.

Weekly meetings of around one hour should bring together the trades people who are delivering work the following week. Each activity is added to the planner and the previous week is assessed for the percentage of work actually completed. Root causes will be identified about why activities were less than 100 per cent complete and learning taken for the following week.

Daily check-ins of fifteen minutes on-site reinforce accountability, allow discussions of problems and road blocks to improve logistics with updates on daily progress. Results of each interaction are recorded and metrics gathered to report on progress and root causes of delays.

The construction industry processes are notoriously inefficient, expending wasteful resources or buffers such as additional time, material inventory and contingencies. Setting targets to reduce time spent by operatives hanging around curtails the space for storing materials, stipulating technologies to drive productivity rates and levels of re-work and defects will push for innovation and radical lean thinking to solve problems. Such targets add greater value to the proposition and offer customers choice between genuine options.

To prompt the radical lean thinking, the supply chain needs to be involved in planning and discussions at an early stage. Collaboration hinges on humans working together to problem solve and that is a cultural shift for traditional construction approaches.

Professor John Oakland is the pre-eminent quality guru in the modern era of Total Quality Management (TQM), statistical process control (SPC) and Total Construction Management. From extensive research, Oakland has set out the behaviours and values for quality professionals in the CQI's publication, *Leading Quality in the 21st Century*, to become quality leaders, who need

> [to] continually promote the real value of quality inside and outside of the business, supporting the delivery of change, building and protecting reputation, and increasing competitive advantage (i.e., cost, time, customer intimacy, service excellence and technical mastery). Those that are successful have worked to be accepted as a truly valuable member of the senior team. This is essential in order to provide a platform of support from the very top of the organisation that is genuinely convinced of the value that the quality leader can bring.[18]

On lean construction, Oakland has highlighted the learning from manufacturing and the TPS in Japan; 'The central ideal of lean production is to give the customer what he/she wants (value) with no waste.' The business improvement methodology is used to identify the waste within and between processes and stripping out non-value-adding tasks and activities.

'Lean quality' has been developed by Oakland as a wider approach than simply product quality and customer satisfaction. It assists in managing the role of business and the supply chain by bringing together different concepts of lean and quality management. By narrowing supply chains and improving relationships, contractors are developing better strategies with the supply chain to lower risk and improve outcomes in terms of cost, quality and programme. Creating an appreciation of the supplier–customer relationships both internally and externally to stakeholders, fosters a service mentality, regardless of whether that is between the contractor and the paying client or the service relationship between a scaffolding team and bricklayers who need the scaffolding for access. Explicitly communicating the customer requirements of the bricklayer to the scaffolder, whether it is working loads or access requirements, builds a better understanding and collaboration between the specialist teams.

Most Tier 1 contractors have demonstrated the potential of lean construction and focused on improving their construction processes and reducing waste (in the broadest sense and not just on site). However, the immense challenge of construction projects in the modern age seems, over the last few decades as projects become more complex, at times to overwhelm the stakeholders and this is why it is time to learn the lessons of traditional quality management and use the available technology to put quality back at the heart of construction management.

Terminal 5 (T5) at Heathrow Airport in London (Figure 4.6) was built using lean construction methodologies. Starting in 2002, sixty contractors were engaged on sixteen major projects costing £4.3 billion that was within budget and completed on 27 March 2008, three days ahead of the 2001 schedule. Some 17,000 tonnes of steel were used in the main terminal building roof and mechanical and electrical (M&E) works incorporated over 1,400 pre-assembled parts using DfMA.[19]

T5, in spite of many twists and turns in the journey to completion, with bag handling problems after launch and, embarrassingly, the cancellation of 300 flights, is generally considered a lean success story with thirty-one million passengers a year in 2017.[20]

With the new economic power it has created, China has arguably become the leader in construction mega projects in the world. The $1 trillion 'One Belt, One Road' initiative to create interlinked connections by road and maritime route is one example of a long list of mega projects. China has thirty-eight nuclear power plants in operation with nineteen more under construction (as of 2017).[21] The new Beijing-Daxing International Airport in the shape of a giant drone, designed by Zaha Hadid, started construction in 2014 and is due to be completed in 2019. It is expected to have eight runways, serving 100 million passengers. The Beijing-Shanghai railway is the world's longest high speed line,

Figure 4.6 Terminal 5 at Heathrow Airport, London, a lean success story completed in 2008.

Source: Belinda Fewings, https://unsplash.com.

costing $35 billion, it was completed in 2011, and is 1,318 kilometres long with 244 bridges and twenty-two tunnels. It used twice as much concrete as the Three Gorges Dam project and is designed to carry up to eighty million passengers per year.

After being discussed for fifty years, the South to North canal system in China finally commenced in 2002. When finished, the Eastern route will be 1,155 km long, carrying 14.8 billion cubic metres of water with the construction of 23 pumping stations and an installed capacity of 453.7 MW. There will be 9 km of tunnels from the Dongping Lake to the Weilin Canal in the first stage. The Central route will be 1,267 km in length and, upon completion in 2030, up to thirteen billion cubic metres of water will flow along it. The Western route on the Qinghai-Tibet Plateau, at up to 5,000 metres above sea level, will take until 2050 to complete, bringing fourteen billion cubic metres of water 500 km from three tributaries of the Yangtze; the Tongtian, Yalong and Dadu rivers to the north-west of China. A total of forty-four billion cubic metres of water will be diverted by 2050, costing around $62 billion.[22] However, it will still leave a huge shortfall from the 200 billion cubic metres[23] needed in the north of China by that year, unless other water projects are conceived.

Notes

1 The White House Building. Retrieved from www.whitehouse.gov/about-the-white-house/the-white-house/

2 Allen, W.C., 'History of slave laborers in the construction of the US Capitol'. Retrieved from https://emancipation.dc.gov/sites/default/files/dc/sites/emancipation/publication/attachments/History_of_Slave_Laborers_in_the_Construction_of_the_US_Capitol.pdf (accessed 1 June 2005).

3 https://vimeo.com/148629834

4 Benning, W., *Code Napoléon* (1804; trans. 1827). Retrieved from http://files.liberty fund.org/files/2353/CivilCode_1566_Bk.pdf

5 Brunel Museum. 'The Thames Tunnel' (2018). Retrieved from www.brunel-museum. org.uk/history/the-thames-tunnel

6 A.J. Downing, who was tragically killed by fire in 1853 on board the *Henry Clay* boat, was the designer behind New York's Central Park, as a radical alternative to a National Mall.

7 Downing, A.J. and Wightwick, G., *Hints to Persons about Building in the Country* (New York, 1847).

8 Fairbairn, W., *On the Application of Cast and Wrought Iron to Building Purposes*. 41st Ed. (1854). Retrieved from https://books.google.co.uk/books?id=ak4OAAAAYAAJ&printsec=frontcover&source=gbs_ge_summary_r&cad=0#v=onepage&q&f=false

9 Ransome, F., *Patent Paving Stone* (1866) Retrieved from https://books.google.co.uk/books?hl=en&lr=&id=66wQAQAAIAAJ&oi=fnd&pg=PA1&dq=building+quality+inspection&ots=8_-lAY_aPy&sig=cSy6JxSr-psry0CRSIppJetT7oE#v=onepage&q=building%20quality%20inspection&f=false

10 Grace's Guide. *Peter Ellis* (2018). Retrieved from www.gracesguide.co.uk/Peter_Ellis

11 Logan, M., *The Part Taken by Women in American History* (New York: Arno Press, [1912] 1972), p. 297.

12 Emporis. Philadelphia City Hall. Retrieved from www.emporis.com/buildings/117972/philadelphia-city-hall-philadelphia-pa-usa

13 Société d'Exploitation de la tour Eiffel. *Origins and Construction of the Eiffel Tower*. Retrieved from www.toureiffel.paris/en/the-monument/history

14 National History Foundation. '"Our Ural". Kasli cast iron pavilion'. Retrieved from https://nashural.ru/article/istoriya-urala/kaslinskij-chugunnyj-pavilon/ (accessed 23 January 2016).

15 Empire State Realty Trust. 'Empire State Building fact sheet'. Retrieved from www. esbnyc.com/sites/default/files/esb_fact_sheet_4_9_14_4.pdf

16 Wakisaka, T., Furuya, N., Hishikawa, K., *et al*. *Automated Construction System for High-rise Reinforced Concrete Buildings* (2000). Retrieved from www.iaarc.org/publications/fulltext/Automated_construction_system_for_high-rise_reinforced_concrete_buildings. pdf

17 Koskela, L., 'Towards the theory of (lean) construction' (1996). Retrieved from https://pdfs.semanticscholar.org/8e87/bc1a102603e9decedf4bb4650803c90f94e4.pdf

18 Oakland, J. and Turner, M., *Leading Quality in the 21st Century* (London: CQI, 2015). Retrieved from www.quality.org/file/494/download?token=UFcUGvXy

19 Laing O'Rourke. 'Heathrow Terminal 5, London, UK'. Retrieved from www. laingorourke.com/our-projects/all-projects/heathrow-terminal-5.aspx

20 LHR Airports Limited. 'Heathrow facts & figures'. Retrieved from www.heathrow. com/company/company-news-and-information/company-information/facts-and-figures

21 Gil, L., 'How China has become the world's fastest expanding nuclear power producer' (Vienna: IAEA, 2017). Retrieved from www.iaea.org/newscenter/news/how-china-has-become-the-worlds-fastest-expanding-nuclear-power-producer (accessed 25 October 2017).

22 South-to-North Water Diversion Project. Retrieved from www.water-technology.net/projects/south_north/

23 *The Economist.* 'China has built the world's largest water-diversion project', 5 April 2018. Retrieved from www.economist.com/china/2018/04/05/china-has-built-the-worlds-largest-water-diversion-project

5   Modern quality management

Modern quality management has been shaped by the introduction of British and international quality standards, various certification schemes and quality marks, quality 'gurus' and the evolution of construction quality control into quality assurance duties for full-time professionals, creating a huge reservoir of knowledge that is arguably not yet fully exploited and utilised.

Quality standards

The UK government started to take a lead on driving quality standards across industries and the British Standards Institution (granted a Royal Charter in 1929) evolved its Mark Committee into the Quality Assurance Council. A flow of written quality standards started to appear in the UK:

- BS 4778, Quality vocabulary – issued in 1971, defining quality as 'the totality of features and characteristics of a product or service that bear upon its ability to satisfy a given need'.
- BS 4891, A guide to quality assurance – published in 1972.
- BS 5179, divided into three parts: Guide to the Operation and Evaluation of Quality Assurance Systems in 1974.
- In 1978, the Department of Prices and Consumer Protection launched a consultative document, A National Strategy for Quality.
- Throughout the 1970s, the endless high profile quality issues afflicting British industry helped to create an atmosphere for improved quality standards and led to the seminal BS 5750 on Quality Systems, first published in 1979. However, it was very much focused on manufacturing and not construction. As it took off in popularity in other industries, it was firmly labelled 'not applicable' by many in building and construction. Unfortunately, that label has still stuck with a sizeable minority in the industry, over the past decades.
- The Building Services Research and Information Association (BSRIA) report, The Application of BS 5750 to Building Services in 1984 began to explain and fathom how such a standard could be used in construction.
- With the updated BS 5750 being issued in 1987, three years later. CIRIA report number 74, Quality Management in Construction Interpretations of BS

5750 (1987) – *Quality Systems for the Construction Industry*, set out interpretations of how translating manufacturing terminology for use in construction.

Such publications had a very limited success in changing minds in the industry. In 1987, the international quality system standard series of ISO 9000 was introduced, based upon BS 5750, which has since undergone several iterations. Changes in the standards' editions accommodated construction and saw the number of overall world-wide certifications increase steadily from 19,768 in 1998 to 116,672 in 2009 but then falling back to just 65,516 in 2017,[1] perhaps reflecting some of the disillusionment with the lack of perceived value in certification.

In 1989, in the UK, the BS 8000 series (not to be confused with the ISO 8000 series on data quality) started to appear on construction workmanship, covering all aspects of activities from excavation to decorative wall coverings. Additional British and International Standards were added:

- BS 10005:2005 – guidelines for Quality Management plans;
- BS 10006:2003 – Quality Management in projects;
- ISO 10015:1999 Quality Management – guidelines for training that provided useful advice on implementing quality on construction projects.

However, there is still needs to be more practical guidance on best practice in construction quality management.

Quality schemes

The poor quality standards in building, especially housing in the UK, have reached the ears of government ministers with monotonous regularity. In 1938, Lewis Silkin MP asked the Minister of Health, Walter Elliot MP: 'whether having regard to the great evil of jerry-building throughout the country, he is in a position to make a statement as to the efforts made by the building industry to deal with this problem?'[2] Elliot stated that the new National House Builders' Registration Council had been established some two years ago previously to deal with the problem.

In 1952, Harold Macmillan, the Minister of Housing and Local Government (before he became Prime Minister), was petitioned by Stan Awbery MP: 'when granting building licences, to make it a condition that the completed buildings must pass the test of a qualified building inspector when they are ready for occupation.'[3] Macmillan assured him that the existence of the National House Builders Registration Council provided the answer.

Adam Hunter MP in 1967 challenged the Minister Dr J. Dickson Mabon to state: 'what progress has been made, or is being made, in protecting the private house purchaser from shoddy building.'[4] Once again, the respective Minister stated the National House Builders Registration Council provided sufficient protection.

In 1977, Sir James Alexander Kilfedder MP urged the Minister Raymond Carter to take action: 'if he will introduce legislation to ensure that those house builders who fail to build to a reasonable standard must make good the defects which develop in a building within a certain period from the date of construction.'[5] Suffice to say, the Minister did not accept there was a significant problem, given the existence of the National House Building Council (having changed its name in 1973 from the National House Builders Registration Council). And so it seems to rumble on, many complaints against some builders and the quality of their work.

A parliamentary briefing paper, in 2018, made clear the ongoing frustration that quality was still not being systematically achieved in new build housing, with too many defects and complaints being reported.[6] The paper also highlighted in the report by the All Party Parliamentary Group for Excellence in the Built Environment (APPGEBE), 'More homes, fewer complaints',[7] which slammed the industry. They state:

- It is an area where we have elected to shine a spotlight because it was clear to us that there is a quality gap between customer demands and industry delivery.
- … 93 per cent of buyers report problems to their builders – and of these, 35 per cent report eleven or more problems.
- There is a perceived flaw in the system of checking quality and workmanship.
- … a decline in customer satisfaction with their new home from 90 per cent to 86 per cent in 2015. That equates to around 15,500 homebuyers (extrapolated from the number of private home completions in 2015) that were not satisfied.
- Another key issue around quality is the so-called performance gap. As many witnesses told us, a gap exists between the designed and the as-built energy performance of new homes.
- Some of those giving evidence pointed to the need for more on-site inspections by independent organisations, in order to drive up quality.

Undoubtedly building quality standards would be even worse if it were not for the ten-year National House Building Council (NHBC) warranty scheme and their 914,000 inspections each year. Given it helps to resolve thousands of disputes per year and paid out £84.8 million in claims in 2017, it demonstrates the importance of a 'backstop' housing quality control. However, it also shows that what should be relatively straightforward production of typically standard housing units of construction, still struggle today to meet basic quality standards.

The British Standards Institution (BSI) was recognised by the British government as the sole guardian for issuing national standards in 1942. The 'kitemark' was duly born and the first one was issued for copper pipe fittings in 1945 (and still issued today). The BSI organised a Commonwealth Standards

Conference a year later, which led to the founding of the International Organization for Standardization (ISO).

The British Research Station (later to become the British Research Establishment or BRE) had a long pedigree in developing an understanding and publishing results on defects in buildings and materials. Between 1956 and 1958,
the BRS carried out inspections of thirteen reputable buildings to better understand craftsmanship and how standards could be achieved. The report was published as *National Building Studies Special Report 33: A Qualitative Study of Some
Buildings in the London Area*. It gave the definition of quality as 'intrinsically not
a subject for measurement, and difficult to define except in terms of durability,
serviceableness, good weathering properties which enhance appearance, and
freedom from heavy maintenance'.

Perhaps there was a centre of gravity argument towards reliability and continuity and even aesthetic appeal, as opposed to measurably demonstrating conformance towards specification but it was a hard look at whether the practical
construction standards achieved the desired outcomes. The report duly noted
that specific written British Standards for the thirteen buildings were rarely referenced. It listed the advantages of using Codes of Practice and explored the
innovative designs and construction especially in the De La Ward Pavilion in
Bexhill-on-Sea and highlighted a perennial problem of record keeping, which
made future changes and repairs more complex. It recommended that records
should be kept safely and future building changes noted, and warned that if
record keeping did not improve, then it would cause ongoing quality issues and
increased costs.[8] These warnings written sixty years ago still ring true today and
emphasise the criticality of Building Information Modelling (BIM).

The National Council for Quality and Reliability in 1962 was created by the
British Productivity Council, and financed by the national government and
organised conferences and produced publications on quality. Industry did not
really embrace such thinking.

Nevertheless, the Ministry of Public Building and Works, the Committee on
Agrément, identified the need for a national authority on testing and assessing
building materials and methods, which triggered the creation of the Agrément
Board in 1966 (later the British Board of Agrément).

In 1968, Sir Eric Mensforth was tasked by the Ministry of Technology with
setting up a committee to look at how quality could be improved in the
engineering industries. Some 800 industry, government and research organisations were consulted. One of the key recommendations was the formation of a
National Quality Board to lead on quality assurance but industry was again less
than enthusiastic and this idea was not developed.

In West Germany, the building control quality system that had existed at the
level between the Länder (states) since 1951 had become very complex and the
Institut für Bautechnik was founded in 1968, based on an agreement between
the State and the federal authorities. In France, the Centre Sciéntifique et
Technique du Bâtiment (CSTB) had an even longer history, commencing in
1947, and was responsible for assessments and testing of new building products

and techniques, similar to both the British BRE and BBA (British Board of Agrément).

In the UK, the Federation of Master Builders (FMB) membership requires financial checks and other company and director due diligence.[9] The only quality assessment comes with an independent check on works in progress. 'Trustmark', 'Checkatrade', 'Trustatrader' and 'Approvedtrades' are just some of the schemes that UK trade companies could join but it can leave customers rather bewildered as to which membership confers the best guarantee of quality work. There are also trade associations ranging from the National Federation of Builders and Guild of Builders and Contractors to dozens of specific trades, such as the Association of Fencing Industries, the British Constructional Steelwork Association, the Glass and Glazing Federation and the Timber Trade Federation, whose membership criteria indicate a level of protection for the customer against 'cowboy' companies.

However, taken in the round, there is minimal proof of the actual quality of work produced by companies who hold such certificates of membership. Now, there will be some excellent companies with expert individuals but the problem for the customer, whether that customer is a consumer wanting a garden wall or a Tier 1 contractor needing major project concreting, is the lack of defined criteria for just quality of work (and not financial stability, sustainability, honesty or other so-called 'quality' criteria).

The inquiry into the Grenfell Tower fire, which was built in London in 1974, revealed appalling design, material and process safety failings, triggering a public inquiry that found various design and quality root causes; no central sprinkler system, only one staircase, missing fire doors and cheap, dangerous cladding.[10] The fire on 14 June 2017 killed seventy-two people and the ongoing inquiry will investigate 'quality of the workmanship' of work carried out on the building over the years.

In 2018, the Royal Institute of British Architects (RIBA), the Royal Institution of Chartered Surveyors (RICS) and the Chartered Institute of Building (CIOB) launched the 'Building in Quality' initiative after the Grenfell Tower tragedy, developing a tracker to monitor so-called quality risks over each of the RIBA Work Stages 0–7.[11] The pack of documents include a basic quality checklist but confusingly includes generic, diverse items such as 'refuse', 'security' and the 'cost plan and financial appraisal'. While being a great initiative in principle, I am left wondering whether the owners had consulted with the Chartered Quality Institute (CQI) and quality professionals. The 'quality' tracker is more a project management tool that includes elements of quality, cost and programming management and fails to appreciate that using the 'quality' should refer to functionality and performance outcomes. Pilot case studies will test the approach through to 2019 and it will be interesting to see if it achieves the aim of measurably improving building quality management.

New quality certifications that extend into the digital environment have begun with a series of BSI Kitemark certifications. In 2016, BSI collaborated with industry stakeholders and launched the BSI Kitemark for PAS 1192–2.

Balfour Beatty Plc, BAM Ireland, BAM Construct UK Ltd, Gammon Construction Ltd, Skanska UK Plc and voestapline Metsec Plc were the first organisations to achieve this certification, 'helping them to build their business in this digital era and access global markets, while giving clients reassurance that they are working with partners at the highest possible standard'.[12]

This BIM Kitemark builds on the verification scheme (PAS 1192–2) and involves the sampling of completed projects, assessment of customer satisfaction through ISO 10004 Customer Satisfaction Guidelines for monitoring and measuring and uses additional assessment parameters through BS 11000 Collaborative Business Relationships (ISO 44001 Collaborative Business Relationships Management System has now replaced BS 11000[13]). Like the verification scheme, the BSI Kitemark for PAS 1192–2 is an important component of BIM Level 2, setting out the requirements for the Design and Construction phase.

Since then, the BSI has also introduced the BSI Kitemark for Asset Management against PAS 1192–3, 'Specification for information management for the operational phase of construction projects using building information modelling'. This demonstrates that asset managers 'have integrated BIM into their asset management processes and confirms that asset information is accurate and up-to-date'.

The BSI Kitemark for BIM Objects was introduced to help validate that the digital version of a manufacturer's product, for example, a window, is an accurate representation of the physical object. Manufacturers are assessed against BS 8541 Library objects for architecture, engineering and construction (Parts 1, 3 and 4) and linked to their quality management system.

In 2018, the BSI Kitemark for IoT Devices was introduced, the first of its kind in the internet of things (IoT) space aimed at the consumer market, and BSI is also developing standards with a number of industries in the digital area, for example, within Connected and autonomous vehicles (CAV).

The PAS 1192–6: 2018, on the collaborative sharing and use of structured Health and Safety information, is a potential method for quality management to be similarly structured and built into BIM models.[14]

The age of the quality gurus

While the quality gurus usually did not work in construction or research, the industry owes them a huge debt of gratitude. Even though construction's productivity has been stagnant for the past few decades, the quality management has undoubtably been better from taking on board the experience, expertise and learnings of W.E. Deming, Joseph M. Juran, Philip B. Crosby, Walter A. Shewhart and Armand V. Feigenbaum in the USA and Kaoru Ishikawa, Taiichi Ohno, Genichi Taguchi, Shigeo Shingo, Noriaki Kano, Masaaki Imai in Japan and John Oakland in the UK, to name but a few.

While the construction industry has accepted some core concepts of quality management in terms of management system, document control and audits (from ISO 9001 standard), inspection and testing (from historical quality control) and just-in-time deliveries (although usually through limited storage space as opposed to JIT philosophy), it has seemingly failed to truly embrace customer satisfaction, quality tools in problem solving, Six Sigma and Kaizen.

Eminent gurus of quality understandably focus on customer satisfaction as the end goal. As an ardent supporter of such philosophy in quality management, it has taken some adjustment for me to shift to a higher goal of lowering risk, especially for safety and well-being. The literature statistically supports the premise that improving quality management in construction directly improves safety, which is of paramount importance.[15] Wanberg *et al.* reviewed thirty-two projects, ranging from $50,000 to $300 million in value. Re-work and defect were used as quality metrics and recordable injury rates for safety. On average, for every 400 hours of re-work, one additional injury was statistically reported.

It is unlikely that Deming said: 'In God we trust. All others bring data' (although he may have mentioned it in a lecture but quoting others) but it highlights the importance of measurement in understanding that data evidence is needed to improve the level of quality performance. Business Intelligence dashboards, when they exist, consist of nominal non-conformance monitoring and rarely do construction companies use genuine quality management dashboards of performance information.

Construction quality professionals

A survey carried out by the Building Research Establishment in 1981 on twenty-seven building sites, covering projects valued then at £100,000 to £12 million, found that nearly all quality control issues were solved between the Clerk of Works and the Site Agent.[16] There were no separate inspector or quality professionals available. The issues covered in the 501 'Quality Related Events', as designated by BRE, in descending order were:

1　Unclear or missing information
2　Lack of care
3　Design will not work
4　Low quality of design
5　No coordination of design
6　Difficult to build
7　Lack of knowledge
8　Poor contractor's organisation
9　Lack of skill
10　Designer not understanding materials

The most successful sites depended upon a positive, consultative atmosphere to problem-solving.

The quality professional in construction today has evolved from the earliest Clerk of Works first mentioned in the thirteenth century in England through to site inspectors of the 1960s and 1970s and today's quality engineers and managers.

There will be a wide variation in specific roles from project to project and company to company. The roles that follow are therefore generic and indicate the shape of responsibilities that may be found in construction contracting businesses.

The quality engineer

The *quality engineer* is predominantly site-based on larger construction projects, working closely with the project manager. During the pre-construction phase, they may develop the Contract Works Information or Specification for quality management of the subcontractors and an outline project quality plan (PQP), setting out quality management roles and responsibilities, how subcontractors and materials will be controlled, inspection and test plans required, non-conformance process, calibration of equipment, document control and special processes. Method statements submitted will be checked and approved, with an eye on how the contractor or subcontractor is controlling the end quality of the work.

Quality engineers will then carry out inspection and tests on site and/or observe and record evidence of compliance with the agreed works specification and method statements. Any non-conformances will be raised with proposed corrective action. Daily surveillance will typically be a walk around the site in the morning advising contractors on any quality management issues, such as delivery and storage of materials, general housekeeping, international and British Standards of workmanship and materials, with either notes being made or a formal site report being produced. In addition, the quality engineer will carry out quality management training and quality tool box talks. A quality engineer will be trained as a quality auditor and will conduct audits according to an agreed schedule and if, non-conformances or issues arise, then they will notify additional audits. They will closely monitor the actioning and close-out of non-conformances. They may be called upon to assist in processing customer complaints, although usually the respective project manager will handle them. Weekly and monthly reports to project managers and the quality managers will summarise findings and make recommendations.

The quality manager

The *quality manager* may manage a number of quality engineers on different projects or may be working alone, depending upon the business size. The quality manager for larger contractors will maintain the quality management systems (QMS) procedures, approve PQPs, develop the annual quality audit programme, liaise with Health and Safety and environmental managers to identify areas of commonality on risk and business assurance, administer the management review (which is often annual) to report on the performance of quality key performance indicators (KPIs) over the past year, develop training and communications initiatives for quality matters and carry out root cause investigations on the non-conformances to identify trends and patterns for improvement of processes.

The head of quality

Quality managers may report to a *head of quality* or head of Safety, Health, Environment and Quality (SHEQ), who will formulate and drive business-wide continual improvement initiatives and advise executives on quality performance. The head of quality will manage the *Quality Manual* and the QMS certification (and re-certification) of the business to ISO 9001 (with a quality manager responsible for the regular dialogue with the certification body on assessment visits). If there is a budget for quality management, then the head will be responsible for the monthly expenditure and financial reports. Customer satisfaction may be monitored through focus groups, interviews and surveys and summaries of complaints duly processed to identify improvements. The head will also maintain the Quality Policy and recommend any changes to the director or CEO, whoever is their line manager. Often a head of quality or SHEQ will report to a Director of Operations or Business Assurance and it is unlikely that those directors will have had much exposure to or experience of quality management. That can be a burden with head of quality battling to be heard among other departmental heads or finding that they have all of the responsibility for the quality management of the business but few levers of power to significantly affect change.

The director of quality

In the rare circumstances that a *director of quality* is appointed, then there is a champion at board level to bend the ear of the CEO and strategically affect the direction of quality management issues by inputting into the business strategy and annual business plan with quality objectives. The director will have overall budgetary and team management responsibilities covering recruitment, coaching, performance appraisals, salaries and bonuses. The director will be the external face of the company on quality issues and may sit on committees of trade associations and professional membership organisations. Sometimes they will be wheeled out to face the media if quality problems impact a project and their duties may include attending and presenting to industry conferences and meetings.

These roles are idealised and often roles overlap due to under-resourcing. The quality team may also include document and information controllers, management system managers and full-time quality auditors.

The sources of quality performance are usually based on the quality audits which are nearly always process-based and not technically based. Red/Amber/Green or 1/2/3 ratings of an audit will indicate if there are major or significant failures in a quality management system, based upon the small samples of evidence seen, but the bias is towards process compliance rather than technical competence in line with business or project objectives.

So, the process may demand that there is a training record kept for Jane Bloggs of her attending a quality awareness course. Providing the record is found and she has attended (often a paper record with a signature of attendance), then the auditor will note down that this is conforming to the process. However, a course attendance is not a certificate of competency. Did she pass the course if there was a test? Was the training course itself relevant to business objectives? Has Jane's quality awareness been measurably improved and applied to her work? Without expertise in training and competency assessments, then a quality auditor's report may skim over and give the impression that training is delivering on quality management education. The audit process can be slow and bureaucratic and the subject matter as a topic may have been randomly chosen six months before on the annual audit programme, to cover an element of the ISO 9001 standard, without any link to current business performance issues. It may well fit with an agreed audit programme covering all the elements of ISO 9001 but may fail to identify pertinent issues that the senior leaders would like to hear about at that time. Not all quality audits are like this but, in my experience, we have slumped into a rut of trotting out activities that were useful in the past and support ISO 9001 certification but do not add real value to a business now.

Quality professionals may have significant influence but rarely any power, which typically resides in the CEO and the directors responsible for design/technical, programme management, commercial, financial and construction areas. Yet each day of every single year of the life of the construction business, there will be significant quality management issues impacting on costs and schedule. A progressive, fit-for-purpose, construction business should have someone on the executive who has a deep understanding of quality management. The CEO should then stand full square behind that individual and trust their judgement. The quality performance needs to be clearly related to business objectives so that the value of quality management can be visibly measured and displayed across the business.

With the technological advances and the fundamental importance of information management, the roles of quality professionals need to adapt. In addition to having advanced digital capabilities (see Chapter 10), they need to develop quality management processes to take advantage of the digital information potential. As Amanda McKay told the author, 'Back in the 1990s, the CQI had a course called "Computing for Quality", which has since ceased but I think we need something like it for today's technology.'[17]

The professional membership organisations, such as the Chartered Quality Institute (CQI), should review their requirements for different categories of membership against those digital capabilities and ensure that they are suitably embedded into their Competency Framework.[18]

Notes

1 ISO, *Survey of Certifications to Management System Standards: Full Results* (Geneva: ISO, 2017). Retrieved from https://isotc.iso.org/livelink/livelink?func=ll&objId=188 08772&objAction=browse&viewType=1

2 UK Parliament, Hansard, vol. 342, col. 594. 'Oral answers to questions: Housing-Building Standards', 1 December (London: HMSO, 1938). Retrieved from https://hansard.parliament.uk/Commons/1938-12-01/debates/1a2ca5b1-e3a1-4fcb-92a7-18b9a9700d83/BuildingStandards?highlight=national%20house%20builders%27%20registration%20council#contribution-7f3d1aaf-7449-44bf-a527-5aeb28a571f1

3 Ibid., vol. 452, col. 195. 'Oral answers to questions: Housing-Building Standards', 4 March (London: HMSO, 1952). Retrieved from https://hansard.parliament.uk/Commons/1952-03-04/debates/2ebaeb17-1023-47f8-8132-82148a497f01/BuildingStandards

4 Ibid., vol. 755, col. 1415. 'Oral answers to questions: Scotland. Private House Building (Standards)', 6 December (London: HMSO, 1967). Retrieved from https://hansard.parliament.uk/Commons/1967-12-06/debates/23b399c4-b8e2-421d-b8ee-300390901fed/PrivateHouseBuilding(Standards)

5 Ibid., vol. 933, col. 1729. 'Oral answers to questions: Northern Ireland. House Building Standards', 23 June (London: HMSO, 1977). Retrieved from https://hansard.parliament.uk/Commons/1977-06-23/debates/9eb4c8b8-63e3-49ae-a37a-54dd823f6af0/HouseBuildingStandards

6 Wilson, W. and Rhodes, C., 'New-build housing: construction defects: issues and solutions (England)' (London: House of Commons Library, 2018). Retrieved from researchbriefings.files.parliament.uk/documents/CBP-7665/CBP-7665.pdf

7 All Party Parliamentary Group for Excellence in the Built Environment (APPGEBE), 'More homes, fewer complaints' (London: TSO, 2016). Retrieved from https://policy.ciob.org/wp-content/uploads/2016/07/APPG-Final-Report-More-Homes-fewer-complaints.pdf

8 British Research Station, *National Building Studies Special Report 33: A Qualitative Study of Some Buildings in the London Area* (Watford: BRE, 1960).

9 FMB, 'Master Builder membership criteria table' (2018). Retrieved from www.fmb.org.uk/about-the-fmb/fmb-master-builder-membership-criteria-table/

10 Grenfell Tower Inquiry. Retrieved from www.grenfelltowerinquiry.org.uk

11 Building in Quality Working Group, *Building in Quality: A Guide to Achieving Quality and Transparency in Design and Construction* (2018). Retrieved from www.architecture.com/-/media/files/client-services/building-in-quality/riba-building-in-quality-guide-to-using-quality-tracker.pdf

12 BSI, Statement: exclusive press release to author, 30 July 2018.

13 BSI, 'BS 11000 has been replaced by ISO 44001 Collaborative Business Relationships Management System' (2017). Retrieved from www.bsigroup.com/en-GB/iso-44001-collaborative-business-relationships/

14 BSI, *PAS 1192–6:2018 Specification for Collaborative Sharing and Use of Structured Health and Safety Information Using BIM* (Milton Keynes: BSI Standards Limited, 2018).

15 Wanberg, J., Harper, C. and Hallowell, M.R., 'Relationship between construction safety and quality performance'. *Journal of Construction Engineering and Management*, 139(2013): 10.

16 Bentley, M.J.C., *Quality Control on Sites* (Watford: BRE, 1981).

17 Amanda McKay, Major Projects Director, Balfour Beatty, interviewed by the author, 17 July 2018.

18 CQI, *The CQI Competency Framework*. Retrieved from www.quality.org/knowledge/cqi-competency-framework

6 Quality information model

The lifeblood of quality management processes is information. Without unique, accurate, timely, complete, accessible, valid and reliable information, then these processes will fail to fully demonstrate performance, i.e. that the as-built product conforms with the design specification and it is 'fit for purpose'. The right information must be collated and supplied to the quality professional, and to all those colleagues who also need quality management information, at the right time that allows decision-makers in business to act.

Digital Quality Management is fundamentally about 'facilitating the performance guarantee' of the built structure and is broken down into the four information management elements required for construction: (1) people; (2) processes; (3) machines; and (4) materials (Figure 6.1).

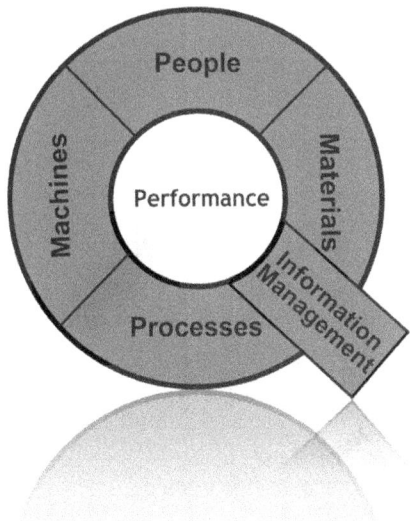

Figure 6.1 Quality Information Model.

Rather than following the usual *modus operandi* of assuming audits, inspections, test reports, non-conformances and other random reports will magically supply the right information, the quality professional needs a business intelligence dashboard to show all real-time information and data that have been summarised and prioritised within an information management process.

This raises the capability standard of quality professionals who need to understand the underlying principles of information management. This may not require specialist qualifications but a close working relationship with Information Technology (IT) and Information Systems (IS) professionals to learn their jargon and firmly keep asking the 'dumb' questions' to translate IT speak into business or quality speak. It also requires quality professionals to clearly map out the information types and sources that they need for both the wider business and specific projects. It may not mean that the quality professional comes to the best decisions or makes the best recommendations (that depends on their capability and specific circumstances), but it increases the probability that they will provide better information to decision-makers.

Information Management becomes a fundamental requirement in the duties of the quality professional. As digital information is not only collated but also may be reported automatically, so the quality professional will find that they will have much more time to analyse and drive continual improvement rather than engage with quality control and fire fighting. That does not mean the roles will be easier but rather will require a different prioritisation. It is likely that this information analysis and continual improvement drive will add much more value to the business and allow measurable improvements to be reported. Overall, this is a positive outcome.

There may be concerns that the quality professional will be stuck behind a desk poring over business intelligence dashboards all day but that should not happen. The quality professional will have more time to travel around, talk to colleagues, to people in the supply chain, to research best practices and propose initiatives that make a difference at the coal face. Intelligent information and better people skills will be the predominant capabilities in future.

Getting alongside designers, engineers, quantity surveyors, commercial managers, and project managers will remove much of the quality control clipboard stigma surrounding audits. Likewise, it will demand relationship-building skills to a higher standard to influence decision-makers in the business and ensure that quality management tasks are embedded and understood within other business processes.

Examples of information that could be evidenced for each performance attribute is shown in Table 6.1. Ideally the information in Table 6.1 will be priority scored but this may be challenging, as, for example, pouring one concrete slab may be different from another slab (e.g. housebuilding to nuclear power plants). In this case, such scoring and prioritisation can still be undertaken by each component within each project management phase.

The Quality Plan (or Design Quality Plan or other Project Quality Plan) should set out the contract-specific requirements for the information management and there must be clear communications with H&S professionals over the contents of the Construction Phase Plan (CPP) and with Environmental

Table 6.1 Evidence of conformance to performance attributes

Performance attribute	Details	Information required: examples
People capability – the extent of someone's ability to achieve an intended result	Evidence of capability to undertake and supervise work	• Competency assessment • Appointment based on qualifications, training and experience • CSCS Operator card
Processes – a set of inter-related activities that use inputs to deliver an intended result	Documented approach to how the work is carried out for end-to-end processes (not just procedures)	• Management Systems UKAS approved certification to ISO 9001 • Process map for concrete pours • Method statement on excavation and setting up the formwork • Risk assessment • Inspection and Test Plans and records
Materials – physical matter and utilities needed to make a structure	Standards of how materials are produced for safety, sustainable use, etc.	• Contract specification • British Standards certification • CE Marking certificate
Machines – construction plant, equipment, tools	Evidence of maintenance from detailed inspections under PUWER down to daily check before use.	• Certificate of inspection • HGV annual test, e.g., mobile cranes and engineering plant

Management professionals over the Environmental Management Plan (EMP). It is best practice to co-ordinate all information requirements with the design team compiled into the BIM Execution Plan (BEP) to ensure that there are no quality management information gaps or duplicate information.

The information could be reviewed as part of a quality audit, but quality auditors tend to focus on just the process management and develop checklists and questions that take process documents, such as procedures, and assess them for conformance of completion of referenced records rather than any meaning-ful assessment of the content of the records. A green-rated audit can be worth little if the technical side of the work is not understood by the audit team. A multi-discipline audit team, depending upon resources, will add greater value and use individual auditors who have full SHEQ multi-capability.

One disadvantage of relying solely on audits is that they may not take place for some time. The audit schedule may be prepared in December for the follow-ing year and although it can be adapted and changed, it is no more than a modest tool in assessing conformance. The evidence unearthed by a year of quality audits tends to be the 'tip of the iceberg' of what information has actu-ally been recorded across the business.

As Mal Stagg, former Skanska Director of BIM, Digital Engineering & Digital Transformation told me:[1]

> Today, a Master Data Management approach should be an essential element of a company's strategy, it needs to be defined and applied at many levels: technology platforms, process/integration implications and the critical behavioural changes required, it is the foundation of good digital business.

Unless a business embraces a full Master Data Management (MDM) strategy, aligned to its business vision, then the naturally fragmented IT will fail to optimise the potential of information management. Traditional data and information were waste products from a final product or service. No one cared what happened to data unless the auditors came sniffing around and then it would need to be dusted off. In contrast, a data strategy places data at the prized heart of a business so that it is efficient to access and share. Agreed protocols mean that everyone knows how to manage data in a standardised way. There is a single source of truth for each piece of data rather than multiple copies across the business. Which is the true version if there are multiple copies? Even with old-fashioned control in the form of wet signatures and authorising names on documents, these can be amended or mistakes can be made. The only way is to be certain is for the true data to be labelled and stored uniquely and then recalled or shared from a fixed location in a database.

Think about Netflix. Customers all access the same version of the film, say, *Jaws*, and it comes from one source. It can be downloaded and viewed offline but it cannot be altered (unless perhaps by some tenacious and technological whizz kids but, regardless, they cannot upload the amended version back to Netflix).

For quality professionals, having copies of data scattered across different applications undermines the ability to interrogate and identify trends and patterns. Demands should be made to package quality management data in logical ways, so that audit report fields relate to benchmarking reports and inspection and test plans. If company names are tagged as 'Business name' in an audit report, as 'Organisation' in an external benchmarking report and as 'Company' in the ITP, then the data user is not interested in excuses by application developers that such data is not relatable and the app developers' conventions need to be remembered by users to find the company in question. It is a failure of internal customer satisfaction to understand that a quality professional needs to assess the company performance across a range of data sources and IT needs to solve these challenges through a data strategy.

Digital Quality Management will create a framework for these records to be captured electronically in a database so that each piece of information on an electronic form is unique and cannot be amended without leaving a digital fingerprint. As an Inspection and Test Plan (ITP) for a concrete slab is completed online, using a tablet rather than a printed form, drop-down menus and refilled information, such as project name, specification reference, pass/fail tolerances, etc., invoke Poka Yoke best practice by making it error-proof or at least reducing the risk of errors being entered into the form.

As information is uploaded to a database, so a picture is being created that should comply with rules and protocols. For example, a process map for pouring concrete will stipulate a rule that an ITP must be completed at a fixed stage in the process. Rules will check automatically that not only an ITP has been created but that the data within it complies with other rules. This allows real-time reporting. Instead of it taking days, weeks or even years to discover a key piece of quality management information is missing or just plain wrong, it can be reported immediately or once other interdependent information is added to the database.

This standardisation of quality management starts to bring discipline and order to an otherwise chaotic world of paper forms being completed that may be inappropriate for the process being undertaken, with fields not fully completed or with data that has '10 cm' being written instead of '10 mm'. A dumb PDF being scanned does little to improve the situation other than provide a back-up copy rather than just a single paper copy liable to misplacement, loss or damage.

By creating such rules once, they can be reapplied over and over to similar work activities with minimal amendments and from project to project. At the beginning it will take patience to identify each piece of quality management information required in an activity or process and then create the rules for checking. Mistakes will be made, information will be missed and results will not be perfect. But that logical approach is far better than the traditional quality assurance approach that fails miserably to provide unique, accurate, timely, complete, accessible, valid and reliable information.

In time, artificial intelligence and machine learning will be able to map out the rules for typical processes, understand interdependencies and even propose changes, according to circumstances. That may still take a few years; not because it cannot be done, but rather who will provide the investment in a notoriously conservative industry? As automation increases, so the need for traditional quality audits will change from a slow, inefficient and ineffective process into one that can report in near real time.

The Quality Information Model is compatible with the 2015 edition of the international standard, ISO 9001 – Quality Management Systems. Section 7.5 Documented Information in ISO 9001 sets out the requirements for information management to comply with the standard, after replacing 'documents' and 'records' from the 2008 version. This requires an organisation to have a holistic view of information rather than simply seeing and thinking in a document-centric world, which is utterly out of date.

Hence, this book echoes the 2015 edition's emphasis on the importance of 'information' and highlights the need to think about it in quality management planning. Given the need for ISO 9001 to cover all industries and reflect a multitude of different business scenarios, then it is hard to envisage this particular standard detailing how information management should be practically implemented for quality management in construction.

Digital quality professionals in 2030

Over time the document-centric world of quality management will erode and almost vanish. Over the coming years, quality professionals need to prepare for the transformation of their duties towards an information-centric world.

A side-by-side comparison of generic duties in Table 6.2 (not specific to any particular quality role) between now and around 2030 shows the dramatic changes coming in quality management duties.

While there will still be a requirement for sound quality management technical knowledge; will it reside primarily with future quality professionals? Given the rapid developments in AI, much of the quality management technical knowledge may be subsumed into AI, making it easier to access by other construction professionals. While the time then spent on educating others on problem-solving tools, such as 8D, 5 Whys, SPC, and so on, may be reduced, it will free up time for other duties, such as quality risk management, or developing new process and performance metrics. Or we may find that there are opportunities for genuinely multi-skilled SHEQ professionals with the business assurance and improvement functions coming together.

Table 6.2 Typical quality management duties

Responsibility	Current: document-centric	Future: AI-centric
Quality Management System	QMS maintenance – *Quality Manual*, procedures, work instructions, forms, templates Project Quality Plans – production and maintenance based on traditional, generic template	Full Integrated Management System with 90% of quality management activities embedded into business processes (rather than stand-alone) PQPs will be built from the BIM model that highlights prioritised quality risks
Auditing	Internal Quality Audits – preparation, audit and issue reports	Audits will be automated to review all digital information in accordance with contract/works information to report on BI dashboard on graded issues found in real time. Remote quality control inspections at suppliers and contractors using AR headsets to walk through premises to assess on-site issues/non-conformances.
Training	Quality Management training – tool box talks, classroom courses	E-learning packages in quality management customised to individuals based on baseline capabilities when they are inducted. Any tool box talks or classroom live training will be through holograms and online interactive gamification presentations.

Responsibility	Current: document-centric	Future: AI-centric
Communications & awareness	Communications – emailed Quality Alerts, newsletter contributions, Intranet articles and web page updates, Yammer updates	Standardised messaging will be linked to AI assessments of topical quality issues that automatically draft suggested communications. These will be reviewed, fine-tuned, approved and published by the quality professional. One message will be published instantly on multiple platforms.
Inspection & Testing	Inspection and Testing – plans' production/approval, create/approve I&T records Equipment calibration – review and approve records Materials – review and approve material schedules, inspect and survey site material deliveries and storage	Generic ITPs for specific processes and materials will be tailored to each design package by AI and reviewed by quality manager. Equipment calibration data automatically streamed and assessed by AI and included in daily update. Laser scanning will replace most inspections of as-built vs design measurements. Videoing by robots and drones will replace witness and hold points. Material inspections and tests will be automated using sensors and results supplied in real time to a BI dashboard
Information Management	Document Control – duties to ensure controlled documents follow version control rules	Skills needed to add a quality management layer of information to BIM models and feed into BEPs. Quality professional responsible for assessing the quality assurance of data flowing through processes.
Project reviews and approvals	Subcontractor PQPs and method statements' approvals as part of a process of project document approvals	Automated process that ties completed work to payment through blockchains. The quality professional may be one person in a chain (e.g. H&S, Env, etc.) whose quality control approvals of work completed will collectively trigger payment.
Problem-solving expertise	Creation of *Handbook of Quality Tools* such as 8D, 5S, FMEA, Pareto, etc.	Comprehensive digital Quality Tool kit accessible through Knowledge Management system linked to IMS. Facilitating crowdsourcing using quality tools will assist with problem-solving.
Continual improvement	Non-conformances register – create NCs, maintain NC register, analyse for root causes	Software will collate design clashes by type and severity. Defects and re-work will be reported online. BI dashboards will collate NC results, identify root causes and offer suggested corrective/preventive actions.

It will depend on how much AI can automate and intelligently analyse and understand health and safety, environmental management and quality management. If much of the drudgery of the traditional quality professional role disappears, then arguably the common SHEQ duties can open up a wider technical capacity for individuals to learn not only SHEQ but also about information security, business continuity and other risk management associated capabilities. Time will tell but quality professionals need to understand that no role in the future can avoid the impact of AI and it is better to shape that future than be swept aside by it.

Quality knowledge management

What have we learnt over the past few thousands of years about quality management in construction? The Great Pyramid of Giza's architect, Hemiunu, the Acropolis' Phidias, the Hagia Sophia's engineer/architects Anthemius and Isidore, the Anji Bridge's stonemason Li Chun, to name but a few brilliant individuals that we know of, did not just design their creations and issue instructions; they were fully-fledged hands-on construction managers who knew the materials, building techniques, and the importance of accurate information and effective communication and had the people skills needed to achieve the quality results they required. They would have walked the construction sites and known the men and women working there. They would have listened and learnt from the generational expertise gathered from the workers and slaves to further enhance their own construction understanding and enlisted them to help solve the day-to-day problems. They would instinctively have known what the client wanted from listening and knowing their patrons' desires (and possibly had an additional incentive of facing the client's wrath if they got it wrong).

Quality control was an inherent and noble skill base essential to building. The knowledge acquired over many, many generations and passed on to young apprentices and trainees established tried-and-trusted methods of working to assess the quality of raw and manufactured materials and components. QC was even more important, given the circumstances of the past. Identifying materials from local sources would save time and money and locating areas for quarrying, tree felling, smelting and building kilns and lime burners required labourers who had knowledge of identifying the best seams of sandstone or the right size oak, to minimise wasting time on imperfect specimens. As we have noted from India, stonemasons were adept at using a metal hammer for testing the sound of a stone for suitability in construction.

There has been a repeated synergy through all cultures of careful thought in creating a built environment in harmony with the natural environment. The Hindu temple building was based upon the rules of the *Vastu Shastras* embracing the *prana* or universal life-force energy, captured in over thirty books written between 3000 BCE and CE 600 in Sanskrit. The *I Ching* book, from the Chinese Confucian culture, has influenced Feng Shui, which seeks to bring humans and nature close together and create a positive effect in living spaces. In the

Japanese writing system, if you take the pictogram (Kanji) for 'house' and the pictogram for 'garden', it gives you 'home'.

The ancient managers supervising material had to know logistics for transporting and what were the best methods for handling and storage of the materials. The issue is not that the ancient cultures had perfect management skills in construction materials, since they no doubt also wasted materials and ordered inadequate types, but rather we should note that they developed an acute sense of the need to understand the strength, texture, colour and durability of the material. We now need a better-trained workforce and a higher level of practical construction knowledge shown by the project decision-makers in the construction business.

These material management skills fall into the discipline of what we call quality management and yet studies indicate that defective materials cost the industry between 5 and 10 per cent of the production cost.[2] With construction contractor profit margins at an average 1.5 per cent,[3] it does not take a genius to question why so little attention is paid to material management quality and the wider scourge of waste in time and processes. We need as a profession to re-develop our appreciation for material management and re-discover a passion for hunting out waste in all its forms.

To embed this quality knowledge, a business knowledge management (KM) strategy should embrace the requirements of quality professionals. As described in the seminal book, *The Knowledge-Creating Company* by Nonaka and Takeuchi, knowledge is the basic unit of analysis, and they set out the difference between explicit knowledge and tacit knowledge in organisations.[4] Explicit knowledge is codified in a written format, such as manuals, specifications, reports, designs and procedures. Tacit knowledge is what happens at the water cooler, when colleagues compare notes verbally, swap ideas and make suggestions. This exchange is not necessarily written down and yet is crucial to any organisation in how it will succeed.

Tacit knowledge can be facilitated through Communities of Practice (CoPs) and Communities of Experience (CoEs) and should be set up to connect quality professionals with others in the same organisation and externally to facilitate an easier exchange of knowledge and wisdom.

CoPs and CoEs can be created by identified individuals with specialist quality management knowledge, e.g. inspection and testing, management systems, data quality, document control, configuration management and laboratory materials testing. A brief summary can be added to a profile page on the business Intranet or using knowledge management software to publicise CoPs. Those leading subject matter experts should proactively set up meetings, presentations, lunch and learn sessions, webinars and other accessible opportunities to both spread knowledge and encourage question and answer sessions.

Quality knowledge can be codified so that it is easier for colleagues to find authoritative texts on key quality subjects. Again, a specialist KM software can be used or an Intranet webpage with hyperlinks. Such signposting should also be comprehensively linked to the Integrated Management System so that there is a 'one-stop shop' for users in how to find information and knowledge. Examples of authoritative texts may include: British and international standards (which may be found through links to specialist databases, such as IHS[5]), trade journal articles, construction case studies, online academic libraries, internal reports and strategies, CQI member resources and books. All such sources of information should be accessed within copyright laws.

The Quality Knowledge Management system may need to be maintained and updated, and this may become the responsibility of the quality professional but this is a great opportunity to keep in touch with innovations and latest editions of standards.

The quality professional needs to develop the capability to teach, coach and mentor others with the aim of improving their colleagues and supply chain stakeholders to become self-sufficient in quality knowledge. 'Quality proofing' the project is an important way of embedding knowledge into day-to-day decision-making when the quality professional is not available and this reduces errors and waste by spreading quality capability across the business.

Self-learning and formal training to become an effective trainer are the attributes of an effective quality professional. Skills to be developed should include classroom-style training, developing e-learning packages, one-to-one coaching and mentoring and delivering stand-up Quality Tool Box Talks (TBTs) on site and presentations to busy executives. The more you practise, the better you become, but these communications are a critical part to winning hearts and minds in the business and gaining supporters and champions who will back additional resourcing for quality management.

Imbuing quality management knowledge into the culture and values of the business is a huge task, especially if the business only currently pays lip service to a basic ISO 9001 certification. However, it is worth developing a Quality KM plan and identifying individuals who are quality enthusiasts to help raise and spread a customer service mentality, as well as harder skills in management systems and quality tools. Next, sounding out the sceptics and even the critics of quality management, can be a useful way to influence their followers and sub-ordinates. Listening to those hard core staff, who may typically have been in the business for many years, talk about their problems, especially with information quality, and establishing ways to improve the processes that affect them most can be an eye-opener, which leads to grudging acceptance and support of other quality management issues.

Getting to the cultural heart of a business takes boundless energy and perse-verance but the value to the business will reap tangible and measurable rewards. The message is 'Don't give up' and 'Keep the faith!'

Digital learning points

Box 6.1 Digital learning points: quality information model

1   Quality information model – identify information on people, processes, machines and materials that leads to required performance.
2   Build quality management data and information requirements into the business Master Data Management (MDM) strategy.
3   Create a Quality Knowledge Management system using CoPs, CoEs, KM software and Intranet webpages.
4   Develop strong quality knowledge skills in training, coaching, mentoring, presentations, site TBTs and wider communications.

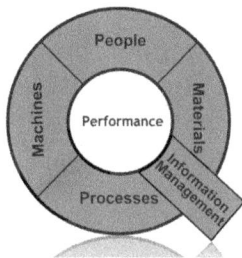

Notes

1   Mal Stagg, former Skanska Director of BIM, Digital Engineering and Digital Transformation, interview with the author, 12 July 2018.
2   Josephson, P.-E., 'Defects and defect costs in construction: A study of seven building projects in Sweden'. Working Paper, Department of Management of Construction and Facilities, Chalmers University of Technology, Gothenburg, Sweden, 1998. Retrieved from http://publications.lib.chalmers.se/records/fulltext/201455/local_201455.pdf
3   The Construction Index, 'Construction pre-tax margins average 1.5%'. Retrieved from www.theconstructionindex.co.uk/news/view/construction-pre-tax-margins-average-15 (accessed 28 August 2017).
4   Nonaka, I. and Takeuchi, H., *The Knowledge-Creating Company: How Japanese Companies Create the Dynamics of Innovation* (Oxford: Oxford University Press, 1998).
5   IHS Markit. Retrieved from www.ihsti.com

7   Data and information management

Data quality

In ancient Egypt, Euclid of Alexandria was a Greek mathematician, known as the 'father of geometry', during the reign of Ptolemy I in the third century BCE. Among his works, he wrote a book called *Dedomenai*, which translates literally from the Latin as 'data'.

Data is the plural of the Latin word *datum*; this is the past tense of the verb *dare* meaning 'to give'. Hence, data has a broad meaning of 'giving' as in when some things are given, then some other things are also given. If we are given something that we know, we can use it in mathematical problem-solving. In the mid-seventeenth century, the word appeared in English in scientific papers and started to develop in its use to refer to principles that were the basis of argument or quotes from scripture that were, at that time, taken as true. By the twentieth century, the word 'data' was used to substantiate the scientific observation from which conclusions could be drawn. In the age of the computer, 'data' began to be used as the term for digitally encoded information, i.e. the '1' and the '0' in binary form that could be stored and processed by computers.

When the history of data is written, many writers will start with Euclid. While he had a major impact on geometry and ultimately the process of information underlying scientific observation, it is often forgotten that human information systems go back thousands of years.

In 1969, Denise Schmandt-Beset started a fellowship at the Radcliffe Institute, in Massachusetts, researching the use of clay objects. She happened to stumble across small clay artefacts in geometric shapes that were part of museum collections across Iraq, Iran, Syria, Turkey and Israel. No one knew what they were. She assembled a vast body of evidence from over 10,000 'tokens', dating back to 7500 BCE, that were an encoding system of accounting data for agricultural products.[1]

The tokens were hardened clay cones, discs, tetrahedrons, spheres, cylinders, ovoids and other shapes that recorded the units of goods received or dispensed. An ovoid could indicate a jar of oil and a cone could indicate grain. In the beginning, perhaps a dozen shapes existed but as the system grew in sophistication, so the number of tokens grew to over 350 in number by 3500 BCE. Shapes included tools and furniture and were increasingly more difficult to make. While a cone was used for a small basket of grain, a sphere was used for a large basket of grain.

Instead of relying on oral descriptions that were easier to alter or manipulate, the tokens presented a physical presence held in the hand. It meant that the actual goods were not always necessary to see or move around as the tokens were reliable representations. A budget for a festival could be calculated by adding, subtracting, multiplying and dividing using the tokens. They prompted new cognitive skills in how data could be encoded and decoded. It meant that the actual goods of oil, grain or clothes could be abstracted from time. You no longer needed to see the actual goods in front of you in order to trade. Hence, a harvest still growing in the fields could be sold or traded in advance of actually cutting the wheat. For 4,500 years the tokens were used across the Near East to communicate the agricultural accounting system.

Clay envelopes started to appear to hold these tokens, with symbols on the envelopes to indicate the content inside, such as three small ovoid shapes to indicate three jars of oil. By punctuating, on the clay envelopes, one dot for a sphere symbolised a '1' and a tiny wedge shape became a '10'. Hence twenty-two jars of oil could be written in wet clay as two dots, two incised wedges and an ovoid outline.

Bureaucratic systems have a tendency to grow more complex and sure enough by 3000 BCE, the authorities were requiring the name of the recipients on the envelopes. Hence, phonograms were created representing sounds.

In the oral Sumerian language a drawing of a man meant the sound 'lu' and a drawing of a mouth was 'ka'. By hence drawing a simple man and mouth together, lu-ka was used for the name Lucas. These phonograms became the first writings of numerals and later became letters in alphabets. It was no doubt soon realised that bulky clay envelopes were not needed and they became small tablets of clay with the evolving symbols etched into them. Hence, writing was born around 3100 BCE.

Writing enhanced the oral tradition of passing on information by means of a longer-lasting and more reliable record than human memory but remembering the original definition of 'data' as giving should be the litmus test when recording it. Does it actually 'give' anything useful? If not, the question should be, do we need it? Arguably we are now drowning in data and it will only become more of a challenge as traditional construction continues to morph into digital construction.

Today Big Data (Figure 7.1) is analysing large sets of data to identify patterns, associations and trends to usefully 'give' insight to stakeholders.

Facts and data are terms that sometimes are confused. When a fact is proven to be incorrect, it stops being a fact. However, when data is proven to be incorrect, it still remains data, as such data is conveying or presenting something that informs.

Typically data becomes information when it has been analysed and can be used in decision-making. Information in turn becomes knowledge when it is

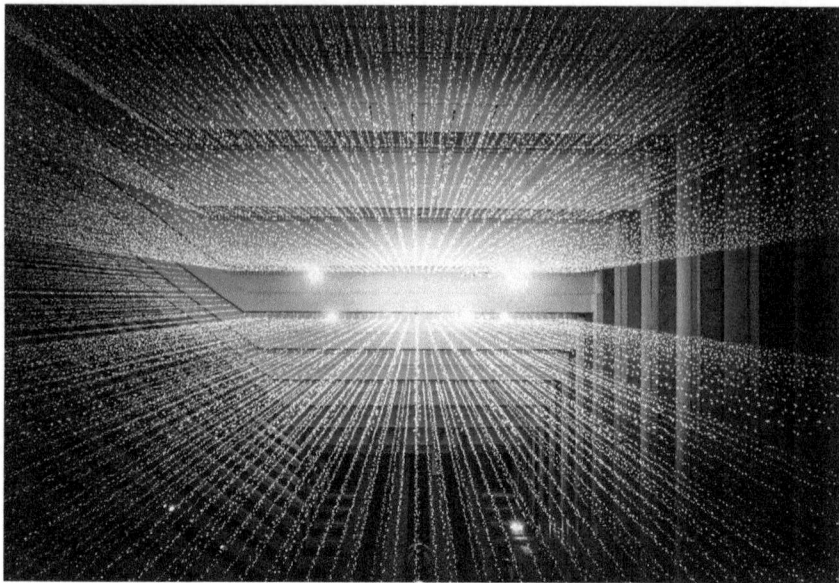

Figure 7.1 Some 90 per cent of data is unstructured and data quality can have a massive impact on creating non-conformances.

Source: Joshua Sortino, https://unsplash.com.

used in the context of experience. Knowledge becomes wisdom when someone knows the best way to apply knowledge. Information Technology is typically used in connection with computers but in this book it is used in its widest sense to describe any technology that can convey information, whether a pen and paper, photograph or spoken artificial intelligence.

The purpose of quality management is to implement the quality policy of the business and achieve the built performance, as set out by the client in the design and specifications. To demonstrate that the required performance is achieved, information is necessary. The information may be derived from data, images and observations in narrative form. Hence, information is fundamental to successful quality management. Yet very often, the information requirements of quality professionals are not adequately thought out at the beginning of a project.

By taking the performance quality attributes (although it could easily be applied to financial or programming attributes), such as the functionality of a concrete structure, the information needed to prove that the concrete meets the specification must be clearly set down.

To identify the quality attributes, the quality professional may suggest to the site engineer (or whoever is delegated site construction responsibility), a generic inspection and test plan for a reinforced concrete slab that will develop the quality information requirements. Table 7.1 provides a generic example.

Table 7.1 Generic concrete floor slab Inspection and Test Plan (ITP)

Project: Oakland University		ITP Ref: XX-OU-F1-005			BIM Ref: Ch-2-J-03		Rev: 1.0
ITP Title: Reinforced concrete ground floor slab							
Quality Performance attribute	What typically can go wrong	Information IN	Inspection and/or Test Responsibility 1 – Subcontractor 2 – Main contractor 3 – Client Conduct test	Witness test	Create record	Acceptance criteria	Information OUT
Strength	Inconsistent or less than required strength	Contract specification	1 – Embed SmartRock wireless sensors and collate data	2 – if required	1 – sensor data	Check concrete specification	Digital QC checklist completed
Durability	Environmental damage over time – cracks, spalling, etc.	Contract specification	1 – Laboratory Water Permeability test on concrete cubes	N/A	1 – Laboratory data	Check concrete specification	Digital test record
Accuracy	Slab not level	Contract specification	1 – Laser scan point cloud imported into BIM model for clash/deviation detection	2 – if required	1 – updated BIM model	Zero deviation	Clash/deviation detection record
	Slab setting out incorrect	Contract specification	1 – Laser scan point cloud imported into BIM model for clash/deviation detection	2 – if required	1 – updated BIM model	Zero deviation	Clash/deviation detection record
	Reinforcement incorrect	Contract specification	1 – Laser scan point cloud imported into BIM model for clash/deviation detection	2 – if required	1 – updated BIM model	Zero deviation	Clash/deviation detection record
Gas-resistant and waterproof	Damage to membranes during installation	Contract specification	1 – Visual Inspection prior to closing up	2 – if required	1 – updated BIM model	Zero visible damage	Digital QC checklist completed
Thermal performance	Incorrect insulation type Poor jointing between sheets.	Contract specification	1 – Visual Inspection prior to closing up	2 – if required	1 – updated BIM model	Zero visible damage	Digital QC checklist completed

For completeness and ease of use on site, additional performance criteria should be added covering safety, e.g. fire resistance and environmental requirements, e.g. sustainability of materials, Albedo Effect, lifecycle CO_2 emissions and thermal mass, to name but a few.

The ISO 9001:2015 standard only makes vague references to data in the sections on processes, customer property and analysis. The ISO guidance on document information[2] also seems to struggle to appreciate the full impact and importance of information management. Hence, most quality professionals other than QA software specialists will struggle to understand how to 'evaluate appropriate data and information' (ISO 9001, Section 9.1.3).

Where I would question the effectiveness of the current edition of ISO 9001:2015 is that it does not seem to fully appreciate that data quality is essential to documented information. How the underlying data for drone GIS data is captured and managed is critical to the quality of information ultimately presented visually in a 3D fly-through representation of a BIM model. If the data has not been quality assured and managed, then it may be incompatible with the digital model or it may even be corrupted or fail to be transferred to the required tolerances. In any of these scenarios, the risk is that the drone operator can do a fine job in capturing the data but their QMS may not adequately protect it in storage or transfer to the contractor or client. The onus is then on adequate arrangements to control the data quality upon receipt by the contractor or client. At this stage, it is very rare for a quality professional to be involved and it usually falls to an operative in the IT Department, who may in fact be quite defensive if asked questions on the data efficacy. That defensiveness is usually based on their lack of training or experience in understanding how to assess data quality and the lack of a wider understanding of quality management. Likewise, some of the defensiveness may emanate from the approach that the wider business should be responsible for their own data quality rather than IT. The best way to solve the issue is to collaboratively draw together the business owner of the data, IT and the quality professional to develop a robust approach to managing the processes so that quality assurance of the data is part and parcel of the process, making it clear as to responsibilities. Typical data quality tools, such as parsing, standardisation, and identity matching and resolution, should be applied to the quality control tasks within the processes.

Where data is being mismanaged by users who are making copies or re-manipulating the data over and above what it is originally intended to be used for, then that is outside of the scope of the quality professional. The key is to return to the business processes and let who owns the requirement for the data to take responsibility for its quality. All other uses and abuses by the rest of the business can be flagged as non-conformances, opportunities for improvement or raised as information security issues to be reviewed. The aim is not perfect data quality, since that is unattainable, given the volumes involved and the frequency of change, but rather to understand the data quality level baseline, assess the impact of poor data and devise a pragmatic approach to process management that minimises the risks.

However, where I am particularly frustrated with ISO 9001 is more in its implementation by businesses, with many only interested in the certificate hanging on the wall and failing to adequate seek out how the standard can add real, bottom-line value. Frankly, those executives responsible in undermining ISO 9001, and its cousins ISO 14001 in environmental management and ISO 45001 (formerly OHSAS 18001) in health and safety, are their own worst enemies, since they fundamentally fail to grasp how these management systems can help to solve their daily problems in meeting respective quality, environmental and health and safety business performance requirements.

For the average quality manager in a construction company, this does not mean getting hung up on the intricacies of data quality but rather systematically challenging and exploring processes where data is being used that have a direct impact on the quality performance. Given that design decisions will frequently hinge upon the digital models, then it is reasonable to assume that all data being imported into those models is quality assured and that those responsible for the importing can point to end-to-end processes with adequate control on checking the data quality.

Table 7.2 sets out an example of data quality metrics that the quality professional may ask for when reviewing or auditing a process.

With data growing at around 40 per cent per annum,[3] quality professionals need to understand that a one-off audit will not suffice and that these types of data quality metrics need to be embedded into a BI dashboard so that trends can be monitored and action taken to drive continual improvement in data quality. In addition, with only 22 per cent data calculated to be usefully tagged, there is a need to question whether the data that is being created, copied, transferred and stored is actually of value to that business.

Information quality

The UK has been one of the leading nations to support the adoption and evolution of Business Information Modelling (BIM) standards with the BS 1192

Table 7.2 Data quality measures

Data quality criteria	Example quality metric
Accuracy	Ratio of data to errors recorded
Consistency	A rule that can check that individual data values are consistent with a total. For example, number of staff by department add up to the total number of staff in the business.
Completeness	Percentage of empty data fields vs total number of data fields.
Integrity	Meeting required specification for data validation by means of structured data testing. For example, in transforming data from one date format to another, what percentage of errors are found?
Timeliness	Accessibility and availability of data to meet expectations. Metrics may include delays in days before data was available against written standards.

suite, which has changed to the ISO 19650 series. BIM is fundamentally about collaboration; bringing stakeholders together around a digital model that improves communications and understanding. Having a 'single source of truth' reduces the risks of design errors and misunderstandings. It is not foolproof, but it is a step forward, compared to the document-centric and paper-centric world of traditional construction.

As Adam Box, Director of ClearEdge3D, once explained to me a few years ago; 'Imagine you're building two schools that are identical to one another. Which would be the better built in a quicker time and cheaper cost?' I replied, 'The second one', thinking it was a trick question. He replied:

> You're right. All those complex design problems having to be solved standing around with excavators means that, when you repeat the build, you know what you're doing and that's what adopting a BIM approach can really help with. You build the first school digitally and make all the mistakes on the computer and solve how to build it in the best way and then when you get on to site, you have a much better chance of getting it right.

The digital 'twin' of a BIM model has been a catalyst for change. As they become more important to construction projects and clients begin to appreciate that they need to have a clear understanding of their own information requirements throughout the asset lifecycle and not just the design and construction phases, so the requirements to digitise traditional documents into digital information increase.

BIM models can be quality assured using software such as Solibri, which checks models against 150 rules, following industry standards. Andrew Bellerby, Managing Director at Solibri UK, explained how he saw the digital future developing in construction:

> Solibri is the leader in the Model Checking category and over the next five to ten years, we will use the power of the software, along with recent technological advancements, such as artificial intelligence and machine learning, to develop smarter ways of working through BIM model patterns in the data.[4]

Quality management is dependent upon data, information and knowledge. Whether a quality professional is auditing a procedure or analysing a set of nonconformances through all activities, information is critical to making and conveying decisions. Information is the oil that makes the project wheels turn and quality professionals should be the ones who should be appointed the guardians of information management and not IT professionals.

In a similar way to checking data quality, information quality can be assessed using the CARS checklist (Table 7.3). CARS stands for Credibility, Accuracy, Reasonableness and Support. While being mainly a subjective list, it provides some guidance as to the reliability of information and tests out the information owners as to whether they have thought this through.

Table 7.3 CARS quality assessments of information

Information quality CARS criteria	Example quality assessment
Credibility – author's credentials, quality control evidence	Usually subjective assessment
Accuracy – evidence-based, fact-based, up-to-date	British and international standards, Codes of Practice, scientific papers quoted in trade journals, case studies with client testimonials, conference proceedings, books (from professional publishers)
Reasonableness – objective, no conflict of interest, fair, no obvious bias	Usually subjective assessment
Support – authoritative sources of information	Bibliography or list of sources that can be verified, if required

Where critical information in design is being quoted as meeting a standard, or during a root cause investigation following a quality incident, check to see if the latest version of the standard is correct.

Quality professionals should be given real authority and power by executives. They should be able to stop work, if data or information quality is compromised. They should have spending power to purchase services to improve quality information management. Talking to one managing director of a company integral to quality assurance in BIM solutions, he revealed that virtually none of the purchasers or specifiers of quality management software were quality professionals. Architects and design and BIM managers were typically those who specified it or purchased it but he could not recall quality managers being involved in even recommending software to quality assure the digital models.

The typical construction project documents and records generated for quality may include the traditional formats for the documents and records as shown in Table 7.4.

It is very much a manual process of creating records from standard templates and forms. The same information is having to be repeatedly input or written that could be automated, such as business name, project reference, user name, user role and date/time. The documents may be kept on an Intranet document management system but invariably multiple copies exist on desktops, laptops, tablets and other individual computers. Unless the document controller, who may be the quality manager, is incredibly diligent and persuasive, then ensuring that only the current versions of such documents are made available is a huge time investment, and there is no immediate proof of compliance since the quality manager is blindsided to what versions of documents are being used across the business.

The amount of time wasted and the number of potential errors that can be created across a large construction project are unacceptable in the twenty-first century. Quality professionals should be mapping out their information

Table 7.4 Typical quality management documents and records

Document or record	Format
Project Quality Plan project specific information and variations to the business QMS	Word processing doc or PDF
Document control records that track versions of drawings, check sheets and plans	Spreadsheet
Quality Training records – attendance, completion and/or pass lists of attendees	Spreadsheet, Word processing doc or PDF
Inspection and Test Plans	Word processing doc or PDF
Inspection and Test records – customised for specific construction processes and materials	Word processing forms, PDFs, photos and video
Quality Audit records and reports – notifications, questions and check sheets, reports	Word processing doc or PDF
Non-conformances – notifications and reports	Spreadsheet, Word processing doc or PDF
Calibration records for measuring equipment	Word processing doc or PDF
Materials – Technical Data Sheets, Material schedules	Spreadsheet, Word processing doc or PDF
Meetings agendas, minutes, presentations for Contract/Project Reviews, Quality Management Reviews	Word processing doc or PDF and presentation formats such as PPT.
Procurement records – approved supplier list	Spreadsheet

requirements prior to construction and inputting such information into the overall project information requirements.

All forms and templates should be digital, with drop-down menus for fields to Poka Yoke them and avoid as many errors as possible, such as standardising the project references, and automatic dates and logins provide user names. Every piece of information should be stored in a database so that all users can then access, share and analyse the information.

Information needs to be pulled by quality professionals to assess the performance of the design and construction and duly recorded to either prove conformance to specification or non-conformance, requiring action to improve.

In quality auditing, sample evidence is captured to demonstrate conformance to specified processes with findings recorded as non-conformances, best practices, opportunities for improvement or observations.

In inspection routines, witness and hold points are documented and evidence, including photos and videos, may be recorded to demonstrate compliance. Test results will set out quality performance of materials and calibration records to show equipment compliance.

The quality management systems should set out the processes of managing the quality of the final build and capturing quality management information should be prioritised, based upon the impact on the quality of this end result. Yet, too often, the processes that can have the biggest risk for construction, e.g. the supply chain procurement, may have the weakest processes and information management approaches.

Amanda McKay, Quality Director of Balfour Beatty Major Projects, told the author, 'Balfour Beatty has been issuing some site supervisors with tablets, so that, following a laser scan of newly built work, they can see if the quality standards required in the digital model, have been met.' She has also urged greater collaboration across the Tier 1 contractors with more 'mature relationships' with the long supply chains to improve information management.[5]

Supply chain procurement processes are too often based on a due diligence checklist to get onto a vendor approval list. The checks include financial accounts, directors, insurance, health and safety, ISO 9001 certification and umpteen other perfunctory requirements, but if a contractor does not have a particular certificate, then it may fall to the Commercial Manager to decide if they should be approved for a project. Their limited knowledge of QA may result in the quality assurance team being asked to help and they may carry out a desktop or site audit. At this point, unless the contractor is completely unsuitable, then more often than not (especially if they have some sort of track record or testimonial), they may be approved. Sometimes for specialised work or a project in a rural area, there may be little competition and a contractor who does not meet all standard requirements is appointed. While this has increased risk, this may not be particularly problematic if they are supervised and their information carefully assured. However, the chaotic, frenetic nature of the construction industry usually means that there is insufficient supervision and poor quality assurance of their information. The small numbers of quality audits and the fact they are by design using samples of evidence mean the probability is very low of thorough audits being carried out of an ongoing contract, where information outputs are critical to the quality of the as-built end product.

Digital learning points

Box 7.1 Digital learning points: data and information management

1 Identify quality attributes specific to project phases and components and then the quality management information needed to meet performance requirements.
2 Data quality – create and monitor performance measures for data, based on accuracy, consistency, completeness, integrity and timeliness.
3 Information quality – understand how BIM models are developed and create a quality management layer in the models. Create a demand on IT for quality management information requirements across the business and in specific projects.
4 Convert traditional quality management documents into digital information.

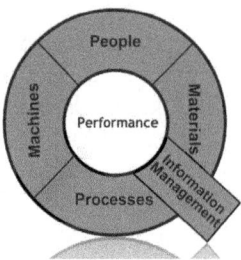

Notes

1 Rudgley, R., *Lost Civilisations of the Stone Age* (London: Arrow Books, 1999), p. 49.
2 ISO/TC 176/SC2/N1286, *Guidance on the Requirements for Documented Information of ISO 9001:2015*. Retrieved from www.iso.org/files/live/sites/isoorg/files/archive/pdf/en/documented_information.pdf
3 IDC, *The Digital Universe of Opportunities* (April 2014). Retrieved from www.emc.com/leadership/digital-universe/2014iview/executive-summary.htm
4 A. Bellerby, interview with the author, 11 July 2018.
5 A. McKay, interview with the author, 17 July 2018.

8 Business intelligence and data trusts

Business intelligence (BI) is simply accessing and analysing information. When I say 'simply' that overstates the $22 billion industry for the software market alone,[1] before adding on the consultancy opportunities. This is the 'Big Data' that many talk about but it often fails to live up to the media hype. Construction is a laggard in comparison to other industries in understanding and using data to improve performance. Construction business leaders need to be enthusiastic supporters of using data and understand the need for a BI strategy and architecture that keeps its eyes on the prize, otherwise it can become a costly, bureaucratic and ultimately useless reporting approach. Each and every director should be the customer for insightful, added value data from IT and its suppliers. Instead, I've seen some directors' eyes glaze over at the mention of data as they skim through the printed monthly report, fixated on some minor issue without coolly seeing the big picture of business objectives, key performance indicators (KPIs) and demanding performance data that tells them how the business performed in the past and performs in the present and the predicted future.

The huge amounts of data in any business can be unearthed and brought to life using Tableau, Microsoft BI, Qlik, Salesforce, Birst and scores of other software solutions.

From mind-numbing spreadsheets filled with endless numbers, decision-makers can see the same numbers visualised into pie charts, line graphs, petal charts, word clouds, scatter plots, timelines and thousands of other imaginative images in glorious technicolour. Figure 8.1 shows simple visualisation that makes data consumption and understanding easier and quicker to digest and thus can facilitate decisions.

While anyone with a reasonable amount of IT knowledge can use a BI software solution and link to simple internal data sources, it is advisable to go to professionals to set up the dashboard. Ensuring data is cleaned up and taking clients through logical process steps before launching into visualisation are vital to ensure that what is seen is true and accurate (although with umpteen caveats, given the vagaries of data management).

There are three key stages to BI implementation. First, it is IT-led, whereby data is extracted, integrated, standardised and organised into data warehouses before applying IT technologies to query and summarise the data. This is a basic

Figure 8.1 Basic business intelligence dashboard visualises data.
Source: @franckinjapan.

'hindsight' approach to BI, looking back into the historical data that is available within a business. It is typically attempting to use past performance to help extrapolate into the murky future, as such, it is loaded with risk. Sometimes the past predicts the future but given the huge changes in technology that impact on business performance nowadays, it may be misleading.

Second, there is the 'here and now' approach, using BI to inform the business of current performance using real-time reporting and alerts. If a performance indicator is being breached, e.g. agreed materials' delivery times exceeded, then an alert will allow decision-makers to reassess the approach and perhaps find a new supplier, thus avoiding the time lag that could affect a construction crew standing around waiting for a delayed delivery. BI real-time data on a fast and furious construction site can make a significant difference as adjustments can be made on a daily or weekly basis to keep up with the programme, but this approach may be rather limited to meeting or improving construction quality standards.

Third, the BI 'prediction' approach supports business modelling and forecasting that can inform recommendations to optimise outcomes. This is the nirvana for business leaders when they are being given reliable options to adapt to new and up and coming circumstances that will allow them to minimise waste (in terms of cost, time, re-work and defects). Seeing into the future is what construction businesses have craved for millennia. It still does not guarantee perfect decision-making by any means but it reduces risk and increases the opportunities for better decisions and outcomes.

Most design and construction businesses are at the first stage of visualising historical data and have a long way to go before this data can provide them with future insights from BI. But before embarking on the journey, the destination should be clearly set out to move into real-time reporting on, for example, the supply chain's ability to deliver prefabricated items to the required quality, using factory sensors, and on to future prediction of, say, quality outcomes of concrete performance before a pour has even taken place. Focusing on the big picture and where the BI project is ultimately heading will minimise the headaches that will arise, if the BI client starts with one add-on after another.

In fact, the biggest headache a decision-maker faces with BI should be; what problem needs to be solved? What business insight is required? It is very easy to quickly start throwing together pretty charts but are they actually adding value? Do they solve pressing business problems?

For the quality professional going back to first principles allows a thoughtful process, by asking the following questions:

- What are the business objectives?
- What are the quality objectives?
- What are the current business problems?
- What are the current quality management data sources?
- Do we need other data sources to visualise the quality problems and monitor the solutions?
- What future insight is required?

Listing current problems and challenges will need input from a range of business decision-makers to shape the most useful charts. That means quality professionals need to understand BI to optimise its potential in quality management.

Quality professionals may be inclined to reach for the old chestnuts of data based upon non-conformances. While such data can be useful in explaining what has previously not conformed to standards, by nature, non-conformances are a lagging indicator and may not forecast what will happen. Twenty-five non-conformances for 'training' over six months for a company with 500 staff does not meaningfully contribute to insight in training. At one company I worked for, the non-conformances categories numbered thirty-seven, and with only small numbers of non-conformances being captured each month, the reports were worse than useless as they potentially misdirected decision-makers to take actions that did not address business priorities. Hours of precious time can be wasted pouring over charts and gleefully pointing to a spike in a line graph or seeing a large 'piece of pie' and then sending dozens of staff spinning off course for the next month, either trying to understand the causes or demanding action. Due to a lack of depth of quality management understanding, the business monthly report contains charts and discussion on these non-conformance 'trends', which create glazed eyes and the report pages are quickly flicked over. It is better to develop KPIs that matter and spend time figuring out the data sources and building an accurate visualisation dashboard, rather than rush to put something under 'Q' and wonder why quality management is not taken seriously by executives.

The data that is most useful is the data that follows the simple rules of, accuracy, consistency, completeness, integrity and timeliness. Sensors can provide performance data in real time. Quality professionals should be robustly demanding that sensors are carefully considered as part of the design process in being placed around the construction, where they can provide useful live streaming of how as-built structures are being constructed.

Location-based sensors, such as GPS (Global Positioning System), RFID (radio frequency identification), UWB (ultra-wideband) and WLAN (wireless local area network), can detect within 1–4 m, depending upon the environment and interference levels, which is acceptable for tracking vehicle routes to estimate material delivery times to site but can be unreliable for activities needing accuracy to a few millimetres, where ZigBee and Ultrasound may prove better options. However, it may well be a combination of these technologies or a new emerging technology that can be a pragmatic winner on site in the future. RFID tags have proved useful in tracking the location of materials, such as reinforcing bars around a site, where the semi-chaotic movement of materials can waste time and divert resources searching for specific materials.

Temperature sensors can monitor shrinkage cracks on curing concrete, avoiding reliance on manual inspection. Pressure sensors can feed back on pre-stressed engineering tests, such as end-bearing capacity in pile foundations. Other light, optical fibre and displacement sensors provide options for data collation. Again, these sensors should be used proactively to solve specific problems rather than with vague notions of wanting to collate any data.

In keeping with a good customer satisfaction business philosophy, dashboards need to be personalised so that the visualised data is relevant to that individual user. In the original set-up, the dashboard should be customised for the user and ideally the user should then be able to continue to change and evolve it, to suit their own circumstances. Permissions can be created to ensure that data security is maintained, appropriate to the user seniority within a business.

Those pretty charts should allow data to be drilled down, opening up other charts and visualisations to provide more detail through subsequent levels. A selection of defect categories could open into sub-categories, project geographical spread, supplier, reportee and year or month.

For the quality professional, it is important to understand the road map of information visualisation to assist in describing to the BI professional what is required. That involves understanding the overall BI architecture and approach to appreciate the context that the quality management BI will lie within, and recognise where there may be synergy, duplication and relationships with other data sets, especially with the other SSHEQ disciplines.

There will inevitably be pushback, scepticism and outright opposition to BI from within a business, from those who are less progressive in their views on technology. They may find fault with data sources providing duff results and may constantly mention that there should be other measures or that results are being skewed. The key is to find common ground with these sceptics, stating that having measures tracked to sound KPIs linked to business objectives is positive and useful for delivering outcomes. Then it is important to mention those measures in meetings, reports, performance reviews and gently scattered throughout daily interactions to demonstrate to staff that these numbers are not for some abstract management report or to tick a box on an audit but an essential tool in the box for management of the business. BI dashboards need to be nurtured and developed but should not become an obsession. They should pop up as part of the personalised log-in to the user interface of the business management system so each team member can see the results and be reminded of performance. It also prompts feedback on the actual measures and if they can be improved, especially in the annual planning exercise. KPIs should be fixed for the year ahead but the BI dashboard can provide a texture to the results in drill-downs and can change if it supports a better understanding of business performance. It may be that an additional data visualisation can improve understanding of a particular KPI or can assist with temporarily investigating a downturn in an aspect of performance. Once the investigation is complete and corrective action taken, then that particular data visualisation may not be needed but should be stored as positive evidence of proactive continual improvement.

The challenge for BI is that, while data can be endlessly sliced and diced and drilled into, the business problems keep shapeshifting. The BI future will be greatly assisted by AI helping to suggest what are the most useful charts to look at, based upon the prioritised problems of the moment.

Practical applications can include real-time updates to search terms within a BI system that uses predictive text to set out possible search terms relevant to the query. The machine learning solution can predict and display automatically the related search phrases. AI can learn from the ever-changing data sources to suggest similar queries and provide better quality results.

Data cleansing by machine learning is another area that will assist in reducing risk in data visualisations as gaps, duplications and errors in data will be identified and either repaired or reported for human intervention. It may be that on a UK BI visualisation for an Inspection and Test Plan, the dates on the data source have been entered in the US format of month first and day second that creates a glitch in the processing. Machine learning should be able to understand such dates without human review. Likewise, a manual data input on a spreadsheet for a survey of spoil heap of dirt on site with one length of 890 metres, hand-measured by a tape measure, which would not fit within a designated area may cross-check with an aerial survey by a drone and be automatically corrected to 89 metres. The number of small but significant errors in everyday construction, that can cumulatively add up to major misjudgements by leaders, will be decreased as AI learns more each and every day.

But further than suggestions of search and automated corrections in data will be creating business systems that use AI throughout different applications to identify trends and proactively create visualisations without someone trying to work out what visualisations they want to see and someone else going away and building the actual visualisation. Automatic BI using AI will revolutionise business decisions, as it seeks out more pertinent data visualisations relating to current business challenges and continually peers into the future to forecast trends and deviations away from target goals. There will be more of a conversation with the data rather than static charts. As the AI learns what the challenges are to the business and relates them to the quality goals, so it will offer up results and predictions to engage with the user.

The typical BI dashboard of coloured charts today may be akin to early children's cartoons that were enthralling in the 1930s, but there is little comparison to today's video games using virtual reality (VR) headsets walking through life-like landscapes in fantasy worlds.

BI may become a constantly changing landscape of business results that you walk through using VR. Walking around a virtual construction site with non-conformances highlighted around material storage depots that produce pop-ups of suggestions for improvements on the level of waste of timber. The AI may highlight, using red triangles in the virtual world, items of risk where a subcontractor with borderline past performance is due to lay a concrete screed. It may automatically adjust quality control with additional witness hold points and additional inspections with instant updates from laser-scanned results.

The VR BI will create its own reports that may be alerted at set intervals to senior decision-makers after the quality professional has added a video or audio introduction or narrative to the VR report. Such future VR BI will be much more stimulating and rewarding to the quality professional in controlling and reporting quality. It will also drive the importance of quality up the corporate agenda with such innovative reporting, which will be rather more interesting than a few pages in a weighty monthly pdf or paper report.

There will, of course, be challenges with the level of trust and confidence in such data visualisations, but given the current problems of manually stitching together data sources, then I would lean towards the AI giving higher quality and more reliable data visualisations than humans.

The wider issues of AI and human engagement then enter the debate. Human intuition is a powerful but often misguided emotional pull, even when faced with overwhelming evidence to the contrary. That will be a key challenge in the twenty-first century for business and wider society to understand and accommodate.

Data trusts

One of the fundamental problems that have dogged the construction industry, probably going back to the days of the Great Wall of China, is the long fragmented supply chain that creates great complexity and inevitably difficult communications. Think of the huge number of meetings that then rely on basic minutes and individuals' recall of actions and events, the ping-pong chains of emails (with attachments) rattling between multiple parties of every kind of issue, contracts, reports and specifications, often posted or as emailed pdfs, typically unrecorded telephone calls, throw in the odd mobile text and paper letters and that is before the digital data of BIM models and business intelligence performance reports, and is it any wonder that the construction industry suffers from misunderstandings and communication failures?

Even the digital platforms using those BIM models as a rallying point for focusing minds and driving collaboration face the perennial challenge that vast quantities of data are configured in different formats and spread across many locations. As such, the models are a basic necessity for complex construction projects but the underlying data of information requirements faces the same web of data cast to the four winds.

Data trusts (also known as data safe havens) have been developed in other industries, notably in health care, as a single location for capturing, accessing and sharing data by stakeholders. It opens up new possibilities for cataloguing and even standardising project construction data so that it is available in more easily retrievable ways between construction project partners, but it also provides a platform to develop smarter methods of analysing data.

The UK government's White Paper on AI set out the case for data trusts:[2]

> To use data for AI in a specific area, data holders and users currently come together, on a case by case basis, to agree terms that meet their mutual needs and interests. To enable this to be done more easily and frequently, it is proposed to develop terms and mechanisms for these parties to form, between them, individual 'data trusts' to enable AI to be developed to meet the needs of the parties involved and allow data transactions to proceed with confidence and trust.
>
> These trusts are not a legal entity or institution, but rather a set of relationships underpinned by a repeatable framework, compliant with parties' obligations, to share data in a fair, safe and equitable way.

Currently contractors, suppliers and clients will take huge numbers of photos and increasingly more videos and privately store them, as evidence of construction activities. That evidence may be for positive reasons of completing work within agreed timescales and to agreed quality standards or conversely as evidence of substandard work. Often the photos and videos are not orientated in space so their geo-locations are not exact but when they are taken and labelled in the right way, they can become a significant and potentially very useful data set. Rather than store them privately, if they are stored in a safe, secure cloud with agreed permission levels for those who have authority to upload and access them, then, by pooling such images, they can be repeatedly used in perpetuity for both short-term analysis on a specific project and long-term repeated use on many projects. Simple artificial intelligence models can use labelled concrete images of good and bad results to relatively quickly identify substandard concrete over newly poured expanses that may take much longer to identify by manual methods of waiting for concrete cube test results and/or visual inspection. The increase in accuracy and speed can be attained by drones cruising over the concrete, videoing, with the AI analysing the images. The more examples added to the data trust of labelled concrete images, the increase in accuracy of assessments.

Issues such as security need to be carefully considered to decide how data can be stored and accessed, but similar issues face all industries and with due care and priority, then such challenges can be overcome.

The architecture of a data trust should be such that it is designed with clear, unambiguous categories and can be maintained at a reasonable cost with free access wherever possible. Some sectors, such as nuclear energy, may need some sensitive categories fire-walled on national security grounds but most data should be accessible throughout the supply chain to encourage buy-in and innovation, regardless of the size of the business.

However, as with all technological solutions, they must ideally be created and used to solve pre-agreed problems. While uses may materialise when a new technology emerges, such as with drones that early adopters have found to add considerable value to flying and capturing bird's eye images of construction sites, data trusts need to start with clear problems to solve to allow the architecture to be designed in the most effective way.

Data trusts will, over time, assist in the development of stable benchmarking of quality standards, so that the industry may permanently develop transparent and quantitative levels of performance. In turn, this will assist in driving continual improvement and innovation across the industry.

For example, performance may be calculated and reported on quality levels attained for pouring concrete or erecting steelwork within specified tolerances. A huge spectrum of data sets on productivity, safety, environment, programme times and materials can be analysed using AI algorithms and reports. Such reports will need to be carefully created to minimise bias and errors but there is a strong potential for pulling down the data that has been uploaded from past construction projects and analysing it to identify best practices and, just as importantly, seek to avoid past failures through using the vast quantities of data already captured within the industry. Moving forward, best quality data can be captured and stored to improve the accuracy and usefulness of legacy data, especially if it is captured with a clear problem in mind.

Regulators and trade associations may wish to become the guardians for protecting specific data trusts to provide independent services and monitor the quality of data. The industry, though, needs to cover the structural costs of maintaining and securing the platforms.

Digital learning points

Box 8.1 Digital learning points: business intelligence and data trusts

1   Influence business objectives and develop quality objectives, KPIs and metrics from them to add value to the business.
2   Learn and understand the BI software solutions – Tableau, Microsoft BI, Qlik, Salesforce and Birst.
3   Use data visualisation to showcase quality measurement. Explain to business intelligence/IT teams the quality requirements.
4   Ensure BI processes are mapped out as part of the integrated management system (IMS).
5   Experiment with data sources for quality – sensors, plant telemetry and radio-frequency identification (RFI) tags on materials.
6   Create and collaborate with data trusts.

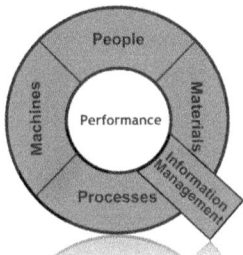

Notes

1 Moore, S., 'Gartner says worldwide business intelligence and analytics market to reach $18.3 billion in 2017'. Retrieved from www.gartner.com/newsroom/id/3612617 (accessed 17 February 2017).
2 Hall, W. and Pesenti, J., 'Growing the Artificial Intelligence industry in the UK' (2017). Retrieved from https://assets.publishing.service.gov.uk/government/uploads/system/uploads/attachment_data/file/652097/Growing_the_artificial_intelligence_industry_in_the_UK.pdf

9 Quality management culture and governance

In the general sense, what do we mean by corporate governance and hence quality management governance? 'Governance' is defined as a way an organisation is managed.[1] The objective of corporate governance can be defined as facilitating effective, entrepreneurial and prudent management that can deliver long-term success.[2]

Best practice corporate governance includes three key ingredients:

- decision-making structures
- processes
- collaboration.

Unfortunately, while businesses have board, audit and remuneration decision-making committees, and will have procedures written down in varying degrees of usefulness, they do not always understand end-to-end processes and rarely develop best practice quality reporting nor high levels of collaboration to enable best practice active governance throughout the organisation.

With thousands of micro decisions being made every day in a business, simply thinking governance relies upon board meetings and auditors' report findings will create complacency and opaqueness in decisions. How many times have we heard the answer to the question: 'Why was that decision made?' is 'Oh, that was Joe Bloggs but he's left the business, so I'm not sure.' There should be clearly defined roles, responsibilities, not just in-post profiles but in responsible-accountable-consulted-informed (RACI) charts for the development of each process and project. Audit trails should hold individuals to account in how risks were managed. Collaboration should be achieved through the use of technology, including BIM digital models as a single source of design truth, effective communications and a positive, open culture to encourage continual improvement.

The ethical side of good governance should run throughout an organisation. Avoiding and declaring conflicts of interest, maintaining confidentiality, good conduct and appropriate use of privileged information are not limited to directors; that extends to everyone in organisations when they are taking decisions.

The aim of quality management governance can be defined as 'facilitating performance outcomes throughout the design, construction and operation of the built environment, by achieving defined quality standards and continual improvement'. The organisation should live and breathe good governance on quality management issues.

The typical form of governance for managing quality is a Quality Management Committee (QMC), chaired by a Director/Head of Quality reporting through to a board director (if the Director of Quality does not exist or does not sit on the Board). In addition, there may be an Audit or Risk Committee that will have inputs of quality audit and nonconformance reports. Other alternatives include a business improvement or business excellence committee that may likewise receive similar inputs.

Whoever is responsible for quality management at the board level may have a short quality report presented to them on a monthly basis, typically containing nonconformance data on numbers received and closed out. There may be some quality KPIs that are monitored but the data will usually be lagging, looking backwards, and will reflect what non-conformances or opportunities for improvement have been identified and reported, rather than any widespread data on quality performance. The report data may be buried within a wider SHEQ monthly report, which will form a chapter of the monthly report of the business, covering finance, business development, HR, design and construction and other areas.

On an annual basis, a Management Review (MR) report may be put together by the quality professional aiming to demonstrate the veracity of the quality policy and that the annual quality objectives have been achieved. The MR report will run through key performance issues on audits, continual improvement, ISO 9001 surveillance audits and certification, customer satisfaction and the QMS. Invariably these reports are in a text format and presented to the board director responsible at the formal Management Review meeting. I have personally sat through such an annual MR meeting that lasted just 36 minutes in total with the CEO busily typing throughout on his mobile, without barely looking up once. The minutes of such MR meetings though, will no doubt record, in great detail, the commitment of the executives to the QMS, continual improvement, etc. to satisfy the brief questions of a certification body.

All of the above are reasons why quality management is brought into disrepute and can be merely a sop to satisfy ISO 9001 certification, which ticks a very small box on project bids. It is extremely frustrating for the quality professional as they can devote a large amount of their time in setting up, managing and monitoring reporting systems, creating reports and making recommendations and administering meetings, all to little avail since the business leadership fails to take any notice and has little interest in addressing the root causes of lower than required quality performance.

A quality management governance philosophy (Figure 9.1) should be embedded in the organisation to link together the business processes, a collaborative culture and decision structures, where it is more likely to bring about the desired

Understand quality risks

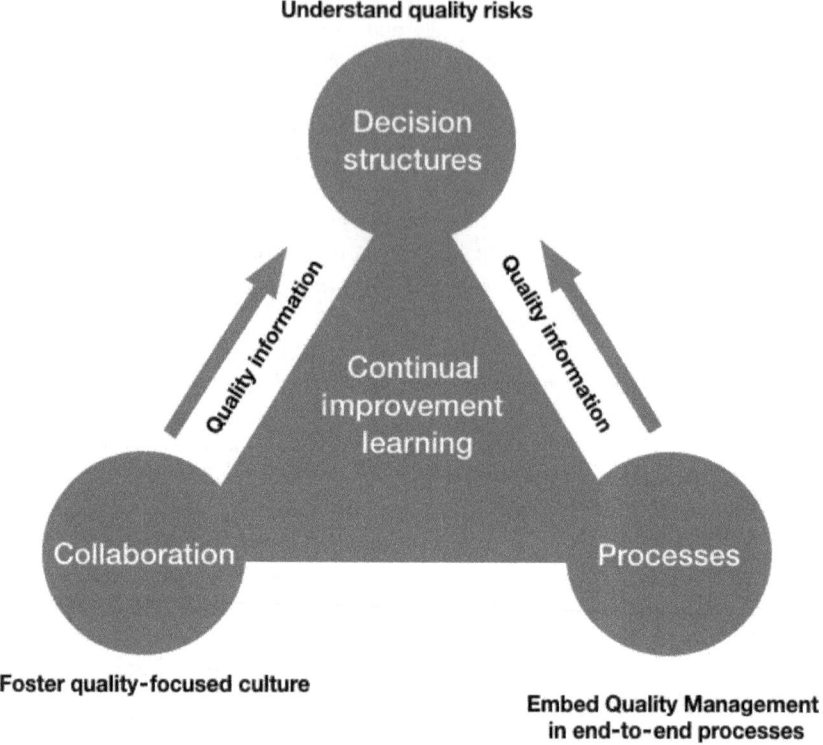

Figure 9.1 Quality management governance philosophy.

results of continual improvement with the right leadership. Of critical import-ance is to have the right stream of information flowing from the process and the collaborative culture to assimilate a better understanding of the quality risks within a governance context.

In projects, assessing quality management information should mean having process metrics and KPIs that the project management team monitor. Having a bespoke BI project dashboard with quality purposely displayed on the front page of the business Intranet and allowing drill-downs into the data, will encourage users to keep quality at the forefront of their minds in the working day.

Ensuring that such data linked to quality objectives is prominently reviewed at regular project meetings and reported up to senior executives, who in turn drill back down and query those reports, will encourage more sunlight to appear in the day-to-day, quality performance of that project.

Business culture can be defined as the 'spirit of the people'[3] working in the organisation or simply its 'soul' or its 'DNA'. The common component in such definitions is the people and how they interact with each other to achieve results.

Developing a new culture of openness is critical to quality management governance success. Unless people throughout an organisation feel that they can safely challenge deficiencies in processes (whether a process needs to be improved or people not adhering to a process), report less than positive attitudes to quality management, report non-conformances and enthusiastically promote quality every single day, then a organisational culture will naturally coalesce around costs and time drivers. In other words, while quality professionals can bemoan the typical performance indicators concentrating on costs and programme, unless we promote quality as part of the organisational culture, then the vacuum will be filled with endless chatter on other business objectives.

This quality culture again needs to demonstrably come from the top and, every day, each executive should be both setting a personal example of quality through praise for work completed to the required level, raising non-conformances as positive learning, supporting campaigns, publishing blog posts on quality, being seen to publicly welcome customer feedback and querying technology issues, such as data quality, and asking how AI can be used to raise quality management knowledge and best practice. Executive advocacy is a powerful signal to staff, consultants, suppliers and subcontractors that quality is of exceptional importance.

Quality management information being fed through to decision-making structures can come in various forms. Measuring quality in staff surveys can provide a measure of the attitudes to quality and allow further follow-ups through interviews and informal discussions in meetings to understand the staff 'temperature' on quality issues. Rather than just a heavyweight once-a-year staff survey where quality can get squeezed down to one or two questions, it is better to use short, sharp online surveys taking less than a minute on a more frequent basis using SurveyMonkey.[4] By using the survey results and adding, say, a 30-minute, interactive 'lunch and learn' over free coffee and sandwiches and an executive Q&A (executives ask the questions to get a sense of the issues uncovered in the survey), a rapid report over a few days can be created on quality cultural issues. The top results should be fed back with a 'you spoke, we listened and here are the improvements' attitude to complete the improvement loop, both raising confidence in the business that quality issues are acted upon and reinforcing a positive approach to continual improvement.

From the results, a 'river' survey map can be produced by process, team, department or other categories to highlight the differences in perceived quality management. By averaging across the scored responses by category and hence anonymising, a map can be plotted of perceived high and lows to point towards sources of best practice within an organisation.

For example, a survey sent to staff states:

On a scale of 1 to 5, where 1 is unacceptable, 3 is minimal quality and 5 is best in class in the construction industry, please score:

Question 1 – The quality of the outcomes for the People Process of Recruitment.

The responses can be collected and an overall average score of 3.6 is established. The HR team have collectively scored the question the highest at 4.1 and the Site Management Team collectively scored it the lowest at 2.3.

Similarly other questions are asked and scored and can be weighted by the numbers of staff in each team:

> Question 2 – The commitment to quality by the Executive Leadership Team.
> Question 3 – The build quality on site.
> Question 4 – The effectiveness of version control by the design team.
> Question 5 – The success of the internal customer-supplier ethos.

The survey questions may focus on just one process or one aspect of quality or one team, dependent upon the results required to support the required quality management and business objectives.

So, with these results shown in Table 9.1, a graph can be plotted to highlight the key differences of perception within the organisation. Figure 9.2 shows how the plotted scores against each question may look for the aggregated scores of only the Site Construction team.

The widest perceptions across the flow of questions are Questions 1 and 2 and, while ideally all question scores should be driven upwards to the highest scores, it is likely to be more efficient in time to concentrate on the scores with the greatest gulf between highest and lowest to make the biggest impact in improvement. Common sense and business priorities will shape such decisions.

With the quality professional acting as facilitator, areas worth exploring for improvement are:

- Q1 – Recruitment perceptions between Site Construction and HR teams.
- Q2 – Executive leadership of quality perceptions between Site Construction and the H&S Team.
- Q3 – Difference in perception between the Site Construction perception of build quality and HR.

Table 9.1 Summary of question scores

Teams	Question 1	Question 2	Question 3	Question 4	Question 5
Finance	3.9	3.9	3.8	3.7	2.9
HR	4.1	3.7	3.6	3.6	3.0
Quality	3.8	3.0	4.2	3.3	2.8
Site Construction	2.3	2.5	4.5	3.4	3.1
Design	3.7	2.4	3.7	4.0	3.1
Security	3.5	4.0	3.9	3.5	3.0
H&S	3.4	4.3	4.0	3.7	3.5

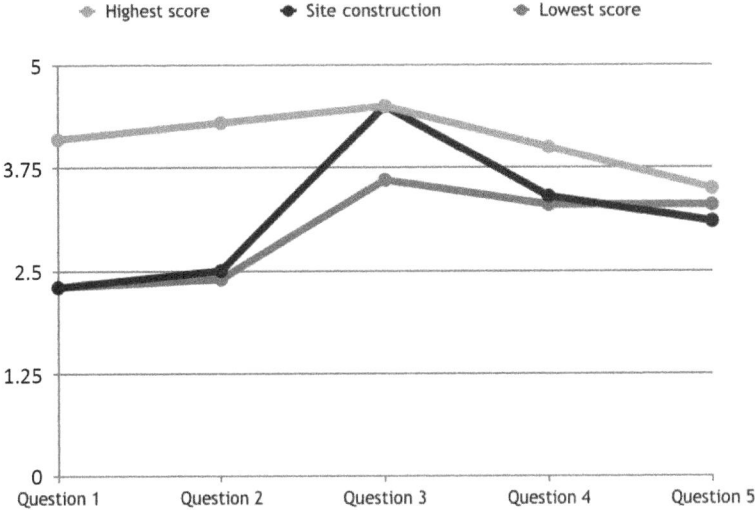

Figure 9.2 Site Construction team scores vs highest and lowest question scores flowing like a river between two banks.

The concept of the river survey map can be used in a variety of different ways such as testing quality management knowledge between teams and using the highest scores to help educate the lowest scorers. Teasing out why one team has a higher level of knowledge in the presence of the lowest scorers can reveal all kinds of improvements from IT access issues to differing management styles on time commitments to CoPs.

A best practice, inverted, internal quality management governance structure is shown in Figure 9.3. This model should be adapted according to the business size, type and culture and some committees may have different names. In small and medium enterprises, the Continual Improvement Committee, Process Development Committee, or Quality Management Committee may become one committee or working group.

The model will have various external quality management checks and balances and stakeholders ranging from government regulations and regulators, certification bodies, CQI CONSig (Construction Special Interest Group) and CQI individual membership criteria, non-government organisations (NGOs), to media and trade journals highlighting quality issues and the public around sites. External complaints should be seen as a positive quality management aspect that gives an opportunity to robustly investigate, improve and demonstrate action to outside bodies and individuals. From industry groups, academia and not-for-profit organisations, the business can benchmark their quality management governance to further improve their approach, methods and risk management.

The *quality management infrastructure* is the expected frontline hard elements to the governance. Continual improvement (CI) and Quality Knowledge

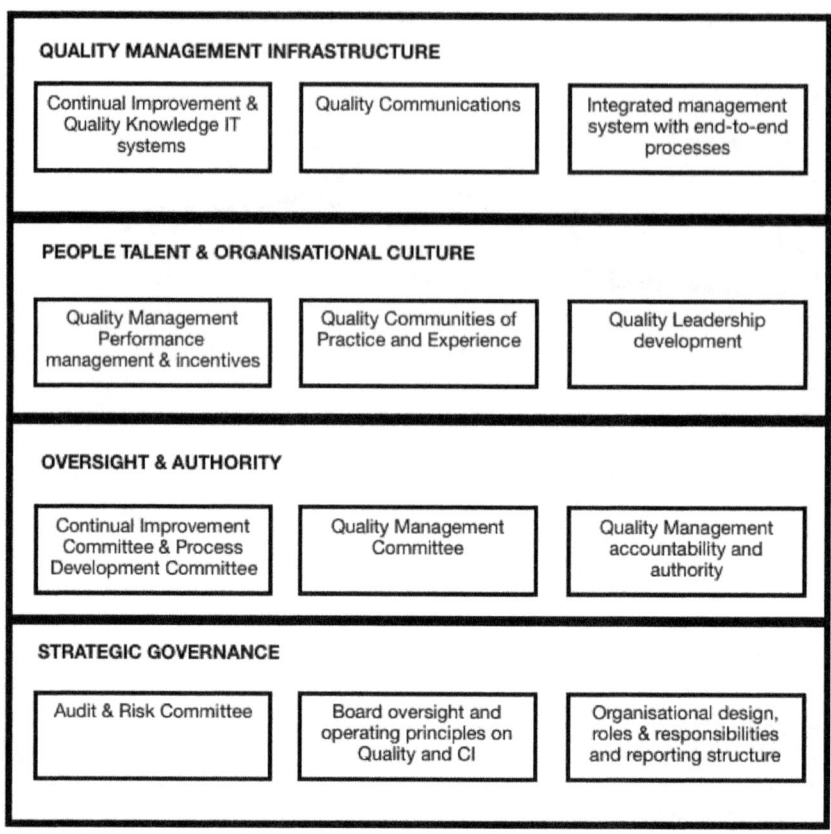

Figure 9.3 Inverted quality management governance structure.

systems will include the receptacles of codified knowledge from databases on non-conformances, to KM best practice case studies that will allow access to guidance and advice in the absence of face-to-face expertise. Quality toolkits are practical tools on problem-solving from simple 5 Whys to more sophisticated Lean Six Sigma and TRIZ. Examples and case studies through e-learning should be easily accessible.

Communications will include staff briefings and quality stand downs, quality alerts, business newsletters, Intranet news bulletins, results of customer surveys, campaigns – activities, posters, videos, presentations and lunch and learn sessions, staff suggestion schemes and social media.

The quality policy, procedures, process maps, work instructions and user manuals, templates and forms should form part of the Integrated Management System (IMS). The IMS and quality KM systems should be closely linked with an interactive, user-friendly portal to encourage use.

People talent and organisational culture form the beating heart of the governance. This is where the preaching needs to turn into practice. Quality management performance and incentives should be based on daily interaction of everyone focused on quality and CI in the tasks they undertake. Managers' rewards and recognition to staff should have quality as a significant outcome with annual bonuses, including quality achieved, as a significant proportion of the bonus. Delegated levels of authority should be published to give everyone confidence to understand their quality 'rights' so that they are empowered to do their job right first time and every time. Inspection and testing should honestly demonstrate to client witnesses how quality is being controlled and should never ever seek to hide defects but rather explain the causes and actions taken to correct the root causes. That will instil far more confidence and trust by the client.

To support the Quality Knowledge Management IT system, Communities of Practice and Communities of Experience are the conduits for accessing listed 'experts' on specific topics, such as document control, data quality or inspection and testing. This may be face-to-face or by telephone or Skype. Regular sessions will be booked and publicised for staff to quickly access those individuals and engage with discussions and debates to solve problems.

Quality leadership is the ambassadorial side to quality management that should manifest itself constantly in people's behaviour. Not just what they say but how they say it. Does the CEO walk around the factory, office or construction site with quality professionals and actively listen, praise and repeat the mantra of quality's importance? Has the intern been given enough encouragement to raise quality issues in meetings and demand action? Are they praised afterwards for speaking up? Is quality leadership evident throughout the diversity of the business to include everyone, regardless of their sex, sexuality, ethnicity, age and disability?

The international standard, ISO 10018, *Guidelines on People Involvement and Competence*, provides useful advice on the process for analysis, planning, implementation and evaluation on developing people capability within an organisation.[5]

Everyone should have access to training, learning and education in quality matters through initial induction, e-learning modules, internal classroom courses, formal education qualifications on quality from CQI Diploma and Certificates in Quality Management,[6] BSI ISO 9001 QMS and auditor courses[7] through to an MSc at the University of Portsmouth.[8] TED-style[9] ten-minute talks are an effective and engaging way for quality professionals to introduce topics.

Oversight and authority bring in the CI (sometimes badged as Business Excellence) Committee with members from a diverse range of staff, focused on weekly non-conformances, best practices and observations raised through a database. The Quality Management Committee should be meeting monthly with at least one director in attendance, who is either chairing the meeting or taking an active role, each time. Day-to-day authority to act on quality should be written into job descriptions or post profiles and the proof of whether it is working is if anyone stops production to address an important quality issue. Progressive companies may issue a credit card-sized, authority-to-act card signed by the CEO, stating that if anyone sees a serious quality management failure, then they should cease work and report/

action it. The Process Development Committee (PDC) will separately assess on a weekly basis new or improved process maps, procedures or other significant management system components. Work instructions, user manuals and forms will be left to the discretion and authority of the process owner to introduce or change. The PDC will have a small number of members pulled from the business assurance functions to provide quality management, environmental, safety and regulatory affairs, with ex-officio experts to be called upon pertinent to that week's system agenda. The process owner will attend to explain the reasons behind changes and will have completed records to show that necessary consultations have taken place and provide a record for any future root cause investigation. The PDC will approve or reject the proposals to maintain process management governance on the management system.

The *strategic governance structures* are shown at the bottom of Figure 9.3, given that by the time they find out there are quality management issues, it may be (too) late and shows that the organisation should be finding and taking corrective action on quality issues where and when they occur. The Audit and Risk Committee can evaluate and monitor quality risks to report to the Board and then the Board should place quality as the top item, after safety. The tone the Board sets on quality management governance is utterly critical. Celebrating achievements and highlighting areas for improvement provide cues to employees that quality is a central part of the business DNA. If the quality is right, then so will safety. The overall design of the organisation and roles and responsibilities should 'hard-bake' effective quality management into the soul of the organisation.

Good quality governance includes open disclosure for businesses. Only publishing a quality policy fails external stakeholders since there is hardly ever any explanation of what the policy means in reality and how it is driving safety and continual improvement in practical ways. Annual reports by business should include several pages on quality management achievements and successes and be transparent about defects, re-work and incidents. These are learning episodes however uncomfortable for the sensitivities of the shareholders.

Quality Management reporting should leverage the technology available for the following:

- BI dashboard live reporting of performance using visualisations that automatically bring out the most important issues.
- VR presentations for executives to walk through sites, offices and supply chain DfMA factories and 'see' incidents of quality challenges.
- Carry out drone inspections during a Board meeting to see issues by video on site, which brings the site quality into the boardroom.
- Apps on quality management governance covering the structure, reporting, ethics of quality compliance, simple communications to raise QMC topic discussions and improvements to post profiles covering quality management responsibilities.
- Microburst online training and e-learning, which may take two minutes and, after a quick test, provide a training credit towards an in-house quality management qualification.

- Short videos should be used extensively for communications. People are more likely to watch a short thirty-second video than read a page of text. Staff should be encouraged and trained in making their own videos to communicate issues by just using their smartphones and tablets. It does not need to be fancy productions, which actually can be counter-productive, but rather homemade videos shot in the office or on site.
- Dry, long reports are ludicrous in the twenty-first century where so much data and information is pushed towards decision-makers that they can miss important trends. Introducing AI to filter, learn and summarise information should be the main way of identifying issues and options. AI can report on levels of staff e-learning and link to levels of non-conformance to identify individual educational packages of topics to learn.

Quality Management audits are one way to evaluate the effectiveness of the QMS through collecting objective evidence to identify risks and to determine if requirements have been met. Internal quality audits are typically undertaken by developing an audit schedule towards the end of the calendar year for the following year by choosing:

- elements from the management system that will represent its core processes;
- aligned or certified management standards, such as ISO 9001;
- management review outcomes;
- past audit findings;
- concerns from stakeholders, such as statutory and regulatory non-conformances, complaints and supply chain issues.

The individual audits may fall into different categories:

- self-assessment by process owners and their delegates;
- desktop reviews of evidence;
- full-on independent audits that include observations of working practices;
- face-to-face discussions and interviews.

Each audit should be carefully planned with questions and prompts researched beforehand, and maintaining the flexibility to chase down issues away from prepared questions in order to follow identified risks. Evidence is noted and non-conformances against the management system, observations, best practices and opportunities to improve can be reported. Multi-skilled audit teams should be considered when high priority audits are being conducted. Ensuring that both processes and technical capabilities are assessed provides a better overall audit than relying on just process assessments.

AI has begun to be introduced for financial and risk auditing and offers an opportunity to improve and speed up the auditing process in the years ahead. Comparing evidence in records with management system processes and recording audio and video evidence taken from operators of processes should allow AI to produce plain English reports that report on measurable findings and the type of non-conformances to areas for improvement. As the AI auditing software will learn and improve, auditor roles will evaporate or have to evolve to a different level of auditing just the AI itself.

These issues lead into quality management AI governance. Much of the AI science requires obscure algorithms to be written and self-learning by the AI that produces results after going through black box thinking, which can be virtually impossible to decipher and prove, without significant resources. There are proposals for an ethical black box to be fitted to robots that will allow operators to ask the AI for explanations of why accidents and incidents have occurred.[10] These will provide additional audit functionality and potentially overcome the concerns of giving AI more and more autonomy to make decisions. Quality auditors may need to be retrained to understand how to assess AI stress-testing routines, overcoming algorithmic bias and interpreting AI outputs for stakeholders, to understand the consequences (although even these more tricky areas will be overcome by AI itself in time). For example, if AI establishes, through statistical process control, the trend analysis in samples of concrete cube strength failures, it is all well and good, but does the variation fall outside the specification and will the client agree to certain concessions, if it is still within certain tolerances? What are the project risks in time and costs of rejecting that batch of concrete?

Digital learning points

Box 9.1 Digital learning points: quality management culture and governance

1   Quality management governance philosophy has three ingredients: decision-making structures, processes and collaboration.
2   Develop a quality culture and benchmark using river survey maps.
3   Quality management governance structure includes the quality management infrastructure, people talent and organisational culture, oversight and authority and strategic governance.
4   Leverage technology into live reporting and drive metrics to decision-makers that add business value.
5   Use microburst videos, apps and VR for innovative quality management learning, customised to different audiences.

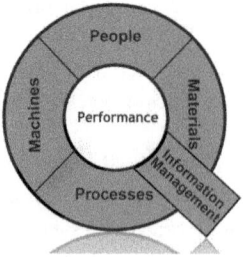

Notes

1 *Collins English Dictionary*, 'Data'. Retrieved from www.collinsdictionary.com/dictionary/english/governance

2 ICAEW, 'What is corporate governance?' Retrieved from www.icaew.com/technical/corporate-governance/uk-corporate-governance/does-corporate-governance-matter

3 kununu, 'What is company culture? 25 business leaders share their own definition'. Blog. Retrieved from https://transparency.kununu.com/leaders-answer-what-is-company-culture/ (accessed 31 March 2017).

4 SurveyMonkey, Retrieved from www.surveymonkey.co.uk

5 BSI, *BS ISO 10018:2012 Quality Management: Guidelines on People Involvement and Competence* (Milton Keynes: BSI Standards Limited, 2012).

6 CQI, 'CQI Training Certificates in Quality Management'. Retrieved from www.quality.org/CQI-training-certificates-in-Quality-Management.

7 BSI, *Training Courses for ISO 9001 Quality Management*. Retrieved from www.bsigroup.com/en-GB/iso-9001-quality-management/iso-9001-training-courses/?creative=194426026494&keyword=iso%209001%20course&matchtype=p&network=g&device=c&gclid=Cj0KCQjwgOzdBRDlARIsAJ6_HNnegv8lMukZ2lDkUzIAtug-hpa07zbY6-ajRuv53lDJGh2eyUBTWYsaAmJrEALw_wcB

8 University of Portsmouth, MSc in Quality Management. Retrieved from www2.port.ac.uk/courses/business-and-management/msc-strategic-quality-management/

9 TED-ideas worth spreading. Retrieved from www.ted.com

10 The *Guardian*, 'Give robots an "ethical black box" to track and explain decisions, say scientists', 19 July 2017. Retrieved from www.theguardian.com/science/2017/jul/19/give-robots-an-ethical-black-box-to-track-and-explain-decisions-say-scientists

10 Digital capabilities

We will call him John, he was a highly experienced quality manager and expert auditor in his field. John was very affable and enjoyed mentoring and coaching less experienced quality professionals. He had many years experience in traditional quality auditing and would pride himself on a disciplined approach. The week before a scheduled audit, he would prepare the checklist and questions before arriving on site and asking for samples of (usually) paper records that would be handed over by the supplier. He would then laboriously write up his audit results and create the audit report the week after and then issue a report that was several pages of text and was rarely read by anyone outside of the team unless it was labelled as a 'red' audit. He was someone I could trust on my team to keep the ball rolling to deliver against the audit programme to tick a box under ISO 9001.

When I proposed that we would be developing live performance reporting from factory floors and construction sites, he was genuinely taken aback, if not appalled. For John, seeing live performance on a business intelligence dashboard, tailored to key decision-makers, of suppliers and subcontractors was an anathema. I couldn't blame him, as he had never come across it before but it showed me the importance of building digital capability within quality professionals. Digital capabilities are skills to a stated competency level that will ensure that the owner lives and breathes digital information and appreciates its value and its potential.

Having a good skill level of information technology in writing reports, basic spreadsheet ability, staid presentations – that is not what this is about. Nor is it super programming skills to create a new application in your lunch hour. Digital capability is a way of thinking about the fundamental importance of information flowing through work and life in general. How do I find out the data, information and knowledge to do my job? How does the organisation treat information; does it have an executed, well-thought-out strategy? How can I use social media platforms to effectively communicate ideas and queries securely and confidentially across the organisation?

Ideally, the education and past experience of the individual will have created a deep appreciation of digital capabilities but that is very unlikely at present and businesses should formally assess the individual's current capabilities and identify areas that can be improved as part of the personal development process.

JISC[1] has an excellent framework for digital capabilities[2] that help prompt further thinking on building essential skills and competency within business. Created for education, the framework is pertinent to the business world. However, regardless of whether the organisation creates a professional approach to digital capabilities, the quality professional needs to develop their own digital skill set to optimise quality management in an organisation. These skills include:

- *ICT proficiency* – using Information and Communication Technology hardware, applications, software and services with the ability to stay up to date with upgrades and new releases. Understanding how to choose the right technology for the right job and working fluently across different digital tools to achieve results. A resilience to deal with simple day-to-day problems (perhaps a bit more than just Control-Alt-Delete …).
- *Information and data literacy* – an intuitive understanding of how to identify, access, manage, organise and share digital information efficiently. An ability to differentiate between the sources of information based on its provenance and appreciating information property rights of copying and manipulating information. Knowing how to interpret data through analysis, raising queries and understanding algorithms and abiding by data protection rules.
- *Digital creation* – knowing how to design and create digital artefacts, such as code, digital writing, audio and video, web pages, apps and the process of production, editing and publishing in alternative work and social media-based environments. Having a drive for innovation through new ideas and knowing how to digitally project manage to achieve results have become more prevalent, as people work remotely when they travel and from home. Digital research to find best practice, new ideas and share evidence are prerequisites for problem-solving.
- *Digital communication* – effective communication in the digital spaces of videoing, phone calls, web chats, Skype, Yammer, email, texts and knowing etiquette, social norms and privacy settings for each media type. Being savvy enough to spot false and fraudulent communications. A critical capability is collaboration through these digital spaces and understanding that sharing information is an established and essential part of improving a team's productivity. Hoarding of information and thinking 'information is power' is a failed and flawed approach and one that must be tackled head-on.
- *Creating and maintaining business social networks* – such as LinkedIn, to build relationships, access information and knowledge are a participation skill that reaps long-term benefits for the business and the individual. Knowing who to go to find useful knowledge is one of the rights of passage in maturing into your employer's organisation. Instead of just turning to a colleague sitting next to you or asking a line manager, going online and tracking down the subject matter expert improves the efficient transfer of information.

It may be the business Intranet or an internal Yammer network that can find the right person, but by doing so, it oils the digital wheels within any organisation.

- *Digital learning* – having a hunger and curiosity to want to digitally learn through self-reflection, being a mentor and a mentee, asking and acting upon digital feedback and efficiently managing one's own time. Being an advocate and a champion for digital learning enriches the culture of a business and makes it safe to share and ask for help. There is so much information and knowledge online that having a disciplined approach to sort through, access and understand what is relevant and useful to the project or task in hand, is a key capability, where there is risk of information overload and it impacting adversely. Knowing the sources of learning from YouTube, Google Books, academic papers, news websites, as well an e-learning for training and qualifications.

- *Digital well-being* – some may recoil from personal, digital 'branding' but it is one of the facts of the twenty-first century that standing out from the crowd, in the right way, benefits both the individual and the business. By developing that stand-out brand of being a SME or the go-to person on specific issues, saves time and improves the flow of information around the organisation. Understanding how to therefore brand oneself with a well-written profile, SMART achievements, or professional photo, across a range of platforms makes you more accessible and approachable.

  Likewise, it is important to protect oneself online and not divulge information that can compromise your real identity. Well-being also covers protecting against the work pressures of being accessible 24/7 and failing to resist every single email pinging into the inbox. Taking responsibility for oneself digitally is important to the business to reduce the risk of stress and anxiety affecting mental health. Knowing how to respond to cyber bullying and digital conflict are skills that are not usually taught and the business has a responsibility to train and make staff aware of these issues and risks.

Telemedicine has been well documented since 1960s and even developed to perform remote surgery on patients in 2001. In a construction project, the fragmented supply chain and assorted stakeholders of the client, designer and main contractor can result in a geographically diverse spread of individuals, sometimes around the globe. For the quality professional, it is a conundrum of wanting to build relationships but not wishing to give up precious time sitting on trains, planes or automobiles, to necessitate face-to-face contact.

It is easy to make commitments of visiting offices, factories or construction sites to see construction activities for yourself, answer queries or help solve the daily problems, but getting the balance right is very hard. As work piles up in a head office in London, a visit to a site on the outskirts of Dundee starts to slip down the calendar priorities.

Setting up quality management online 'surgeries' is one way of providing stakeholders with face time and proactively creates efficient time in the calendar to be more accessible. It cannot replace the real face-to-face meetings, audits and inspections but it builds relationships and engages individuals to pass on quality professional expertise.

Online surgeries are part of the wider practice to be seen 'out and about listening' across the business, supply chain and wider stakeholders. It is a form of branding (however much people may recoil from the idea) but it is important to reach out in a world where there is a growing expectation to socialise within business. Tips are shown in Box 10.1.

In the beginning, it can be a little daunting and yet it will be appreciated by most staff. Heck, it gives them a bit of an excuse to down tools and watch a bit of television! Quality professionals should put out of their minds nervousness of making mistakes; it will happen, but soon it will become part of the business 'furniture' and there will be great interest, suggestions and quiet appreciation.

Conversely, a shambolic presentation with the quality professional not knowing simple digital skills, such as how to share a presentation screen on Skype or muting themselves repeatedly when they shouldn't, will test viewers' and listeners' patience. A few mistakes will be forgivable in the early days but online communication skills and competency need to improve quickly to maintain audiences and credibility. Practice, practice, practice beforehand, that is the key to create a professional online presence and put other people at ease. Some traditionalists will say that quality professionals do not need to be Hollywood actors

---

Box 10.1 Remote online surgery tips

Be seen to be out and about, listening.

- By creating a regular one-hour, monthly surgery in the quality professional calendar, any stakeholder can then call in using Skype or, if Wi-fi is difficult, phoning for audio. Choosing a regular time slot helps to embed the activity in busy people's minds. Getting directors to actively and openly tune in is key to demonstrating that it is important and part of the business culture.
- Invitations can be emailed out to all staff and others can be added to the invite list. Other communication channels, such as e-newsletters, company magazines, Quality Alerts, etc. will over time promote the surgeries.
- It takes practice to feel at ease in front of the laptop camera and unless questions or comments are going to be pre-screened (seriously not to be recommended), then those dialling in may present challenging questions. Be prepared but be honest if the answer isn't known immediately!
- It is advisable to start the surgery with a short five-minute briefing on a quality management topic. Partly to break the ice and partly to use the opportunity to communicate a useful topic.
- Record the surgeries but ensure everyone participating knows at the beginning that a record is being made. That allows issues to be captured without distraction and creates a useful video library available on the company Intranet.

and they are right, but they must be able to deliver engaging and interactive presentations, regardless of the format.

I have seen online surgeries that feature regular presentations using shared screens on Skype, safety messages, competitions and director announcements. Sometimes the surgeries can be very quiet and other times the audience numbers can be very respectable.

By using analytics from Skype, the anonymised viewing figures can be seen and broken down into location and team. It is interesting to see the numbers from sites who participate and giving a shout out to a particular project or site is a good way to bolster interest and appreciation from them, when they may often be overlooked or feel they are.

Online surgeries also allow a riposte to any complaints that the quality manager is not seen enough on site or in a regional office. If based on site, then the quality professional can use the surgery to reach back to the head office staff. Specific surgeries can be created where company issues need to remain confidential. They can be set up just for suppliers and subcontractors or for a select group of client representatives.

They become an invaluable learning source for the quality professional. Where people will be reluctant to report a defect or item of re-work, they will mention it in a surgery and either the quality professional can capture the information or a bit of online encouragement can get the issue recorded there and then by the individual.

Over the following four weeks, issues can be investigated and at the following surgery reported back to strengthen the feeling that 'things get done' by the quality team. The visibility of issues is vital to reinforcing a culture of continual improvement.

Another opportunity for remote online expertise is site operatives arranging an appointment to discuss with the quality professional a site issue, where the site operative uses a camera to visually highlight a problem. It may be plant, materials or process-related but by pointing the camera at the issue, the quality professional can provide better advice and suggestions than just through a telephone conversation. The video cameras, smartphones or wearable cameras on site may be used to shoot live or record an issue and send to the quality professional shortly afterwards.

It also provides another chance to record the issue for re-use at a later date whereby faulty component supplies or a failed inspection allow the quality professional to create a case study or Quality Alert that can be broadcast or uploaded to the business Intranet rapidly for a topical communication that may prevent the same issues arising elsewhere.

The lurid headlines of mass unemployment for certain industrial sectors from robots is great for clicks and artists creating humanoids but it oversimplifies the reality and seeks to sensationalise a trend that has gone on for thousands of years. Namely, technology, whether stone hand axes, printing presses or computers result in some jobs ceasing and new ones being created, as we acquire new knowledge and invent new methods of production. Nevertheless, estimates widely vary but suggest that between 10–30 per cent of jobs in the UK could be automated and between 9–47 per cent in the USA, impacting different industries in different ways.[3] Some research in Germany from the manufacturing industry has suggested that roughly for every new robot, there may be two fewer human jobs but this is *more* than offset by growth of jobs in other sectors.[4]

Construction jobs are estimated to be at the top end of those figures (around 40 per cent) as the technological change moves from automation of simple computational tasks through to automation of repeatable tasks and on to automation of physical labour. Those scary figures fail to differentiate that some construction workers who have a higher level of education, are likely to find little impact (around 4 per cent) on jobs, by 2030.[5] The trick, therefore, is to assist people in education, training and access to self-learning with free online courses, to adapt to the coming technological changes by improving their skills and capabilities. It can be difficult for those undertaking manual jobs now to wonder if they will be affected but rather than 'wait and see', *all* construction workers should be looking to continually improve their knowledge, skills and capabilities. To that end, the UK government has recognised the need to pump prime investment in new skills as stated in its Industrial Strategy with a key policy that will:

> Create a new National Retraining Scheme that supports people to re-skill, beginning with a £64m investment for digital and construction training … [including] a total of £30m will be invested to test the use of artificial intelligence and innovative education technology (edtech) in online digital skills courses so that students can benefit from this emerging technology.[6]

This total was increased to £100 million a year later by the Chancellor[7] and construction businesses should be queuing up to put forward their initiatives and plans to partner with government initiatives. Independent of business training, quality professionals should be acquiring Digital Quality Management capabilities through self-learning and move to become protagonists and advocates for technological change, so that they have a greater chance of increasing their value to organisations and, hence, decrease the chance of being affected by the coming automation in the next few years.

HR, or rather People and Talent departments in businesses, will be faced with supporting the digital transformation of managing people, bots, robots and AI entities working together. Some tasks will require just humans working with humans, other tasks will require just robots or bots working by themselves but sometimes humans will be working 'side by side' with robots and/or bots. Imagine the skills needed to communicate and ensure satisfactory outcomes with bots and robots conversing with humans. If HR thought people working with people was a challenging environment, mixing in artificial workers ups the ante.

New jobs will be emerging that the quality professional will interact and engage with; Augmented Reality Journey Builder, Person-Machine Teaming Manager, Digital Tailor, Quantum Machine Learning Analyst and Master of Edge Computing, may be some of the new roles that will emerge in businesses through to 2030.[8]

The future workplace is starting to emerge with greater transparency, which is a prerequisite for driving collaborative performance. As AI systems become more common and more intelligent as they absorb more data, so they will first support and then partner with HR departments and line managers. Performance management, both in terms of overall individual and team objectives and day-to-day achievement of tasks set by line managers, is composed of smarter measurable activities adding up to the sum of project outcomes.

AI will become better and better at predicting and forecasting project outcomes. At first, it will be simple metrics based on time and money. When will an activity within a project be complete? Based on limited information and data, it may mean that only basic tasks such as a long brick wall or completion of a building roof will be forecast, as it can under some BIM programmes. Then, as it learns from more projects, so all the variables of human design errors, material defects, weather and poor training and sickness levels start to be absorbed, allowing better and better accuracy of forecasts. Not only will current performance then become more transparent but also organisations can be given predictions on project completions and identify down to the individual where there may be drags on performance. AI systems may take on the role of performance management and this may make it uncomfortable for some individuals who do not wish to be measured in small increments and in a more open manner, where other team members can see who is the 'bottleneck'.

This means that the line managers and directors need to be much more capable of understanding those performance drags. There could be a myriad of reasons such as individuals not receiving other information outside of the project in a timely fashion or their own performance may be sub-optimal but for good reason, such as mental health issues.

However, reaching for the 'hire and fire' solution is a failure of line managers and directors. Usually, it is counterproductive, since removing a team member alters the team dynamics and increases stress on the team. Recruiting a new team member results in delays of three to six months, while someone is found, interviewed, gives notice elsewhere, is on-boarded and then has the learning curve of understanding the new role. For projects this has a significant impact on outcomes. If the original recruitment was done well, then root causes need to be diagnosed and solved; information delays outside the project need to be freed up and better support given for team members with mental health issues.

For the quality professional, the new skills required will be designing information processes to collect and interrogate data that provides good human intelligence on the project performance, constantly assessing issues to maintain performance targets. Also appreciating the need both to link up with AI systems and that AI will be continually improving predictions and requiring more data, so that data flows will never be static but rather ever expanding and deepening.

Interacting with a bot will likely require a more standard method of communication, at least in the early days. Typically, business communications between employees is through email, instant messaging, such as Skype, and sometimes business social media apps, such as Yammer. How often do humans add all sorts of trivia, social comments, jokes and jargon that will flaw a bot until it has acquired such understanding? Staff may need to be taught that communicating with a bot requires clear instructions and requests and the social chit chat needs to be left out.

What happens when AI becomes more reliable and better at tasks than humans? Imagine people being made redundant, since AI in bots can measurably do tasks more efficiently. The stress, jealousy and inevitable workplace conflicts will need to be handled carefully and proactively by quality professionals working with HR Departments, who in turn may be partially run by bots! There may be virtual workers and managers who are AI. They could be issuing instructions to people, which means that quality engineers may have a line manager who is AI. Quality reports will be produced in real time as sensors stream data on material performance on site and on components being assembled in factories, which will be analysed and visualised on dashboards on tablets and smartphones, requiring quick decisions.

All these scenarios will start to play out rapidly over the coming years and smart members of the Quality fraternity will need to embrace such changes and challenges with an open mind. Those who recoil and push back, hankering for days of clipboards and writing up audit reports over several days based upon one-month-old evidence will not survive long. Speed and accuracy of AI will require a different kind of collaboration, willingness to trust in artificial intelligence analysis and reporting and making decisions much more quickly.

Digital learning points

Box 10.2 Digital learning points: digital capabilities

- *People capability* – do staff know how to use Skype? Can they access e-learn/IT Helpdesk to find out? Is the quality professional confident enough that they have the capability to undertake online presentations and know how to share screens, record surgeries, troubleshoot, etc.? Can the quality professional create and edit videos for concise reporting of quality issues arising from sites?
- *Processes* – It is worth documenting the process for remote online experts to iron out issues of who does what and when.
- *Machines* – Do all staff have Skype (or equivalent) accessible on their laptops?
- *Materials* – How will presentations/recorded surgeries be made available to staff after broadcast? Are they securely maintained to prevent loss or theft?

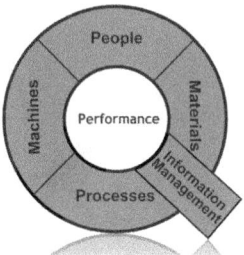

Notes

1 Historically, JISC stood for the UK's Joint Information Systems Committee.
2 JISC, 'Building digital capabilities: The six elements defined digital capability model'. Retrieved from http://repository.jisc.ac.uk/6611/1/JFL0066F_DIGIGAP_MOD_IND_FRAME.PDF
3 Frontier Economics, 'The-impact-of-AI-on-work' (2018), p. 32. Retrieved from https://royalsociety.org/~/media/policy/projects/ai-and-work/frontier-review-the-impact-of-AI-on-work.pdf
4 Dauth, W., Findeisen, S., Südekum. J., and Woessner, B., 'German robots: the impact of industrial robots on workers' (2017). Retrieved from ec.europa.eu/social/BlobServlet?docId=18612&langId=en
5 PriceWaterhouseCoopers. 'Will robots really steal our jobs?' (2018), p. 31, Figure 6.6. Retrieved from www.pwc.co.uk/economic-services/assets/international-impact-of-automation-feb-2018.pdf
6 HM Government. 'Industrial strategy-building: a Britain fit for the future'. White Paper. CM9529. (2017). Retrieved from https://assets.publishing.service.gov.uk/government/uploads/system/uploads/attachment_data/file/664563/industrial-strategy-white-paper-web-ready-version.pdf
7 Ryan, G., 'Hammond pledges £100m for National Retraining Scheme', *Times Educational Supplement*, 1 October 2018. Retrieved from www.tes.com/news/national-retraining-scheme-philip-hammond
8 Cognizant, '21 jobs of the future: a guide to getting and staying employed over the next 10 years'. White Paper (2017). Retrieved from www.cognizant.com/whitepapers/21-jobs-of-the-future-a-guide-to-getting-and-staying-employed-over-the-next-10-years-codex3049.pdf

11 Web-based process management

The principle of a management system is to write what you do, do what you write and then improve. It then allows an auditor to look at the written word and compare this to the real world and through objective evidence, discussion and observation, assess the effectiveness of the work. That is one of the key tenets to developing, using and improving a management system.

Typically, a traditional quality management system will consist of a quality policy, quality manual, procedures, work instructions, guidance and forms to create records. It follows the traditional document-centric format for any part of a business management system (BMS), as shown in Figure 11.1.

The aim is to 'consistently provide products and services that meet customer and applicable statutory and regulatory requirements' and 'to enhance customer satisfaction'.[1]

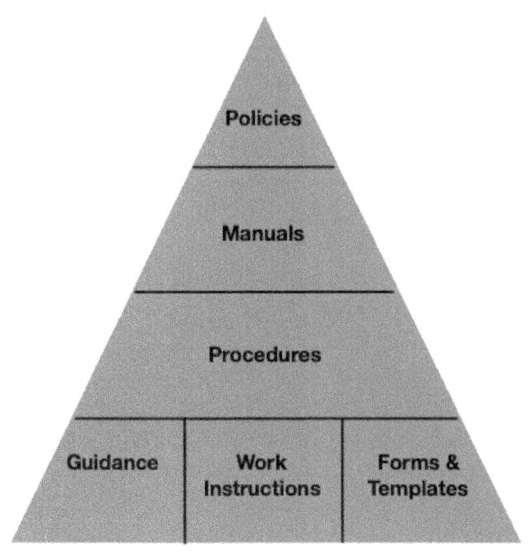

Figure 11.1 Traditional document-centric management system.

The problems with current management systems are:

1   The time lag from writing, to approving and publishing the document can mean that the process has already changed.
2   Authors are not always good writers and poorly written documents find their way into management systems.
3   Humans invariably find workarounds and short cuts to get tasks done, thereby there is often a mismatch between a carefully thought-out best practice and what is actually being done.
4   Management systems that are document-centric have a habit of growing and growing. Rarely do they get pruned, and from a modest set of documents over a few years, they can become monsters with obsolete methods and tasks still remaining and endless, unnecessary detail in other parts.
5   Few people read the published versions of the documents ever again (unless they are preparing for an audit and blow the cobwebs off) and so until process owners review them or auditors assess them, it is not known if they are still current and accurate.
6   Usually, a contractor's management system can be constructed and operated in splendid isolation to other stakeholders. There may be access points where clients, the public and the supply chain interface but these can be an afterthought. It hardly helps to support a seamless, integrated approach to achieving results.

The typically live elements of the management system tend to be forms as they are used regularly and, regardless of what the procedures may say, will typically drive the process behaviour. People develop routines around recording information and data and sending it to someone else or filing it away.

Let's stop for a moment and ask: what is a management system truly for? It sets out information on how, when and why to do a process or task. People need this information but usually turn to colleagues around them to find the same information. It is quicker and easier to speak to someone than to find the right document and read it. Once people capability grows to understand and follow the process or IT workflow software drives behaviour, then usually the written document becomes an awkward afterthought. In the fast-changing, dynamic business world, the reality is that the traditional BMS, and QMS within it, cannot compete with people finding ways to do 'stuff' that do not necessarily follow the best constructed management system.

Where the management system used to be endless pieces of paper printed out with authorised 'wet' signatures by directors who had usually never read (and certainly not digested and fully understood) the documents, so folder systems, like SharePoint were introduced to store the documents electronically in neat files. While this has helped to alleviate the felling of quite so many trees (although people still go round printing off parts of the management system), the interfaces tended to be very basic, 1990s-style façades with a few hyperlinks. It required knowledge of the management system structure to click multiple times to drill down to the right document. Even though these systems have a search engine, unless keywords are regularly updated, then the limited functionality of the search engine is very likely to come back with many hits, for umpteen versions of the same document and may not actually provide a result for the document the user needs.

The 2015 version of the ISO 9001 standard has even permitted doing away with quality management documents, such as manuals and written procedures, leaving it to auditors to assess consistency from evidence gleaned by discussion and observation. While the theory is laudable and it does allow some discipline in processes, it can be weak and ineffective since the time is limited on an audit for auditors to speak to enough staff to assess consistency between them. The problem is not necessarily with the premise of removing written elements to the QMS but rather it makes it even harder for senior managers to know if people are consistently doing what senior managers think they should be doing (until the results of the activities become known). I write this as someone who has spent nearly thirty years in and around management systems and I want them to work.

More effective, modern management systems use a process management architecture to enable a more logical, fully integrated approach to understanding how a business should be managed. There is much confusion over processes and procedures. Processes are steps that have an input with value added and an output that gives an intended result.[2] Process maps (flowcharts, see Figure 11.2)

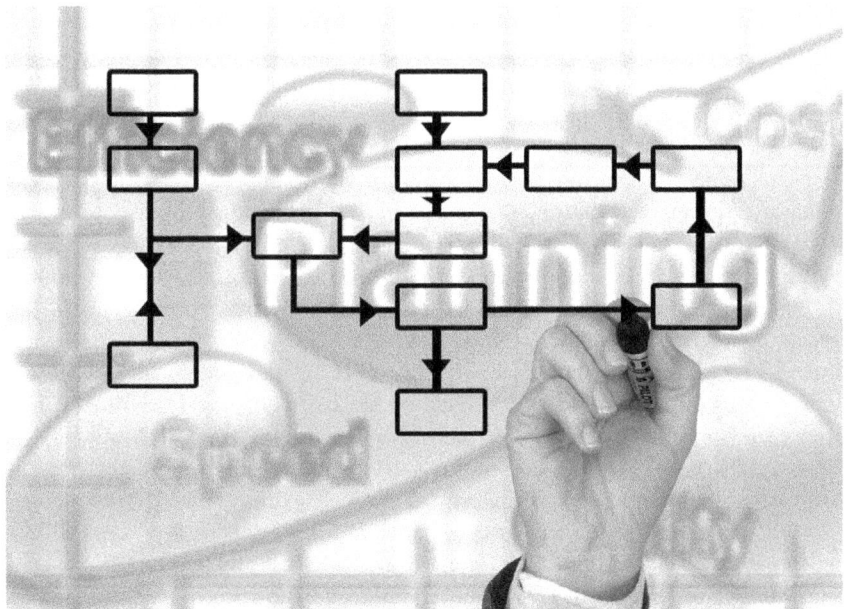

Figure 11.2 A simple process map or flowchart sets out the steps for turning inputs into value added outputs.

Source: Gerd Altmann, https://pixabay.com/en/users/geralt-9301/.

visually show what the steps are in each end-to-end process and are a better way to convey those steps than writing a paragraph of text. A procedure is a specified way of doing something but may or may not show an end-to-end process. It is better to make a clear distinction of keeping a process with simple, steps from start to finish and creating procedures to provide more detail of tasks that add up to part of a process.

The top level of the process architecture will typically fall into three categories for a construction business:

- *management and control processes* – business strategy, governance, risk management, continual improvement, business assurance and regulatory affairs;
- *core processes* – design, configuration, supply chain construction, operation of the asset and decommissioning;
- *support processes* – people, property, knowledge, IT, change, business continuity, communications, data and information.

This top level may be shown as a single-page graphic setting out the overall composition of the integrated management system.

The second level will individually map out all these individual, end-to-end processes, such as business strategy (from management and control processes), design (core) and IT (support).

The third level of processes can be developed from using a SIPOC methodology in workshops with groups of users to pull out for each process,[3] Suppliers-Inputs-Process steps-Outputs-Customers, and hence allow outputs in one process to be linked to inputs in another process and likewise, name suppliers and customers for each process. In this way, an integrated process architecture can be drawn for a business.

Manuals and guidance notes may still be useful to provide context and background to the management system but they should be out of immediate line of sight on the interface, so as to provide only supportive information.

Intuitive user interfaces overcome the problems of older, stagnant interfaces through design from rigorous research and testing to entice users to enter and access documents, either through predictive text word searches and/or hyperlinked hierarchical documents. Good practice suggests technically observing how a focus group of users access existing management systems with the way they search information, the short cuts they use, their pain points and frustrations, and access routes as they journey through the system, and this reveals drop-out rates of finding information, in order to understand the challenges users face and get a feel for what they want from the system. User types will be modelled to establish the way different groups of users – by sex, seniority, technical discipline, disability and other segment audience types – use the system and may have different needs. A prototype interface will be built, tested and improved by the same focus groups before launching.

Moving from one existing management system and rolling out another either has to be undertaken as a 'big bang' overnight or a weekend with sufficient education beforehand and support after the launch, or as a transition where parts of the system are taken out and replaced. The latter can be messy but necessary if there is a risk that by the time the new system is completed, it is already out of date and in effect becomes 'painting the Forth Road Bridge' with never a good time to launch it. Change logs need to be maintained to a greater or less degree, depending upon regulatory and certification requirements and become an audit record for future root cause investigations.

In the past, large bound volumes of paper procedures and manuals would be kept on shelves, if you were in an office or site cabin. On a construction site, though, at best, a clutch of paper forms would be carried to record evidence and so the process needed to be in an operative's head.

Even with the introduction of the laptop and then the tablet, accessing the management system has not become much easier. Given the size of some text documents, even word searches are time-consuming and trying to juggle viewing multiple documents on one or more screens to find the right information is usually deemed more trouble than it is worth.

High risk construction sectors such as rail and nuclear energy take a rigorous approach to documenting and reviewing management system components, but the everyday pressures on process owners still mean they struggle to maintain a disciplined approach to ensuring that the content is constantly relevant and up to date.

One key aspect to management systems in high-risk industries is the quality assurance graded approach that aims to avoid an overly prescriptive system, given the propensity to add layers of very detailed documents within the system. It recognises that not all construction activities carry the same risk and to apply the best controls where there is the greatest risk.

The graded approach will also support the effective use of quality assurance resources, by prioritising activities requiring greater quality management oversight and scrutiny. The quality assurance grade should be directly linked to the safety/security/environmental/procurement grades to prevent confusion and set out an easy to understand framework to choose the most appropriate grade.

Even if the basic steps in a process are maintained, trying to enforce the process on users is another challenge altogether. How does a manager who is juggling technical problems, personnel issues, budgeting, a crisis and the usual daily administration, find time to constantly check that their team is individually each following umpteen activities within umpteen different processes? This is where the new technology begins to disrupt management systems, together with a realisation that day-to-day activities need to be more flexible and adaptive than the typical, bureaucratic management systems allow.

Latest technology on site has evolved to beam augmented reality procedures into smart glasses, so that operatives can access instructions using eye movement and overlay them in their surroundings. By looking at a QR code next to valves, the heads up display (HUD) will overlay the sequence for checking pressure dials and turning taps. Photos can be taken by eye control of the dials without the need to write anything down. That means that an operative does not need to read the manual beforehand in an office and memorise it, before walking to the location or fiddle around on a tablet with freezing hands and bob their heads up and down focusing their eyes first on the screen and then the valves, trying to translate words into actions and back again. Working in this way increases the probability of errors many times, over the course of a year.

Consideration needs to be given to how the client management system upstream and the supply chain management systems downstream connect into the Tier 1 contractor's management system. Typically, there are few, if any, connections thoughtfully set out into the clients' own management system. However, it makes sense to create a framework of connections. The client may require access into parts of the contractors' system and hence be given remote permissions to view-only key parts. A common project management system that goes beyond the usual project plans may be created for each specific project that develops *joint* processes that both parties approve:

- A risk management handling process is best practice to understand not just which risks are identified but how they are managed on an ongoing basis between the client and the contractor.
- Design change management is a crucial process to utilise technology to seamlessly allow design changes to be initiated and controlled.
- The BIM digital models should be the single source of truth in design for all stakeholders to access, with quality management issues highlighted in context to the model.
- A joint conflict resolution process will allow rules of engagement to be approved before disagreements commence, that will aid how conflicts are managed and resolved.

Such jointly approved and followed processes should improve collaboration, trust and engagement between the parties and overall assist in creating positive relationships between individuals. ISO 31000 is a useful reference for understanding principles of risk management.[4]

Project Quality Plans (PQP) flow from the contractor's IMS and should be aligned in format to BS ISO 10005 to set out the 'specification of the actions, responsibilities and associated resources to be applied to a specific project'.[5]

I have seen a number of PQP templates used by contractors that simply state a long list of other documents relating to the contractor's IMS, without specifically setting out how quality is to be achieved on the specific project concerned. It has led to a site manager walking around with a Lever Arch file full of documents, making zero sense as to how quality is being managed. Why reinvent the wheel when the recommended format and structure can be found in an international standard?

All project quality components should be geared towards risk management of how to achieve the specified results and not just trot through mentioning required standards, processes, resources and methods. Likewise, the plan should include how undesirable outcomes will be mitigated.

The biggest challenge facing the use of a PQP based upon best practice, is how it is kept alive. Other than internal audits prompting a review or a quick read through, most of the time the PQP will gather actual or virtual dust, and the owner of the plan will rarely initiate updates, changes and improvements. The key to a successful PQP is taking the required quality controls and outcomes on the project and ensuring that they are embedded into control systems that demand their appearance at appropriate times in processes. While collating all aspects of project quality management into one place is useful, the outcomes need to be embedded into technical processes to optimise the chances of meeting the performance characteristics.

If a concrete drainage channel is being constructed at a certain level, then a laser scan should be part of the inspection and test plan when the earth is being excavated, to ping back the result and automatically trigger an acceptance or rejection within defined tolerances. Where the requirement is embedded in the method statement for that activity, it should be attached to the BIM digital model. In such a way, the quality outcomes are digital and it is easy to check whether they have been achieved and this will also create a digital record for future review.

Developing key performance indicators (KPIs) from the people, process, materials and machine metrics allows the management system to be measured and improved. Process metrics may, for example, measure the number of defects but adding trend characteristics such as by product/day/location can allow tracking that will initiate more effective improvements.

Ideally, a chain of interdependent metrics will be built up so that if the performance metric A meets X standard, then it moves on to metric B further in the process that measures Y standard. An alert will be generated if metric A fails to meets its standard, allowing a quick identification of the problem along the process. Metrics need careful consideration and it may take trial and error to create a useful one that adds value and is not just collecting data for the sake of it.

By measuring the number, owner and type of clash detections in a digital model and adding trend characteristics by day and week, the data may also reveal a pattern based on who is designing and whether they could learn from how they model. Research[6] has suggested that isolated design working, non-BIM-specific training and the current structure of cloud-based common data environments (CDEs),[7] do not easily facilitate clash *avoidance* in BIM. Without measurements, therefore, it is difficult to identify trends and hence potential improvements.

Tracking the number of inspection failures on site with the percentage of bricks damaged per pallets, can illuminate process management issues, for example, on material supplies not being adequately protected in transportation and handling.

Ingraining regular reviews of processes can become habit, if process owners receive friendly but prompt automatic reminders by email or text or a friendly word in the canteen to carry out self-assessments, which can be carried out by desktop, observation or as a mini-audit asking questions and requesting evidence from colleagues. These exercises are a useful way to get to know processes inside out and see the variation that requires action to change behaviour or rewrite the process.

The integrated management system (IMS) will seek to embed as many activities into the end-to-end processes, so that most SHEQ activities are included inside processes owned by process owners other than H&S/Environmental/Quality professionals. While building 'power' is part of organisational politics, by giving away such ownership, it makes the process owners more accountable. They will be seen as a customer who should be proactively asking for SHEQ services, rather than being prodded and pressed to give up time for audits. The 'pull' is a demonstration that process management is in the right hands.

Accessing the IMS should be available through all platforms– desktop, laptop, tablet, smartphone and smart glasses – which means the systems need to be configured so that formatting and layout of the management system remain user-friendly.

In addition, using the wiki concept of www.designingbuildings.co.uk/wiki/Home, works by following how people want to find information on their own terms. Searching for a word or phrase and having hyperlinks throughout the text to information within the same site or to reliable external sources, is user-friendly. It is continually updated and maintained by a community of experts who keep the community informed of changes to standards and regulations. Management systems need to adopt these sort of user-orientated information and knowledge management systems to be relevant and useful.

The Digital Construction Management System (DCMS) in Figure 11.3 embraces the key components of a construction contractor's business: a process architecture, digital information and IT systems.

These components have advantages over current management systems:

1   The process architecture should be robust enough not to have to change much over time. The fundamental components of policies connecting to the suite of tiered processes, management and control processes, core processes and support processes should remain constant. Below these top tier processes, individual end-to-end processes may change but the trick is to minimise these changes and remain vigilant for the architecture to start growing legs.

2   Knowledge management systems need to flow from the process architecture to allow codified, explicit knowledge to be written down and tacit knowledge to flow around an organisation that may or may not be codified at some stage.

3   IT workflow systems should be introduced wherever it is possible to standardise routines and yet still allow the software to be updated and amended as required. Workflow systems, such as Asana, Monday, Workfront, remove the need for separate forms to be created and managed since the information required to perform a task is embedded into the system. For example, a

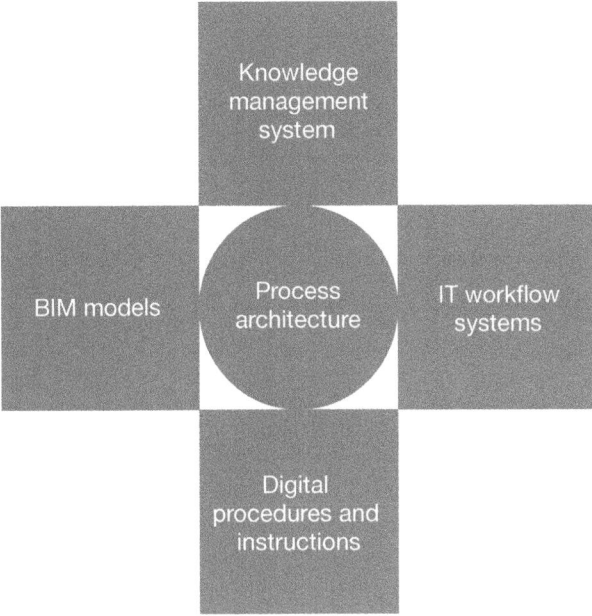

Figure 11.3 Digital Construction Management System (DCMS).

recruitment 'procedure' is driven by the software requiring details of the recruit to be entered and approved at appropriate times according to the process built into the software. No more consulting between procedural text and forms, with the inherent risks of errors or short cuts. The onus, however, is on the workflow being carefully designed and built around user needs and according to best practice. This is where the quality professionals, business process analysts and IT specialists need to work closely together.

4   Work instructions can be added to heads up displays (HUDs) or smart glasses can be introduced for higher risk routines typically on site and prone to mistakes. These allow the user to access instructions explaining what to do without resorting to manual and procedures.

5   BIM models should have layers of information built in on quality management, environmental management, health and safety, security, and so on which drive collaboration around a single source of truth for the virtual built asset. This provides the connection of *what* is being built with the *how* it is being built. Dress rehearsals of 5D BIM models demonstrate the time lapse changes needed to increase people capability of how to construct on site.

The DCMS is a more dynamic approach to managing a business and process owners (POs) need to drive changes and improvements to the overall system.

POs must be highly collaborative by nature and have the authority to overview all routines within their end-to-end processes, whether that is knowledge, workflow, HUD instructions or BIM models. By understanding information flows throughout the business and throughout all processes, the POs can improve their processes quickly and effectively (providing they consult with other POs) without resorting to endless director approvals.

Digital learning points

Box 11.1 Digital learning points: web-based process management

1   Develop a Digital Construction Management System (DCMS) to create a more dynamic approach to integrate knowledge, workflows, HUD instructions and BIM models around a robust process architecture.

2   Develop a user interface for the management system, which is user-friendly, intuitive and easy to navigate.

3   Use SIPOC methodology to develop process maps, starting with top-level processes and working down to the lowest processes. Ensure that all processes connect with outputs flowing into inputs to build a true IMS.

4   Embed hyperlinks in the IMS to connect to quality knowledge management systems.

5   Use workflow software for processes where it makes life easier for the users and pares down procedural text.

6   Investigate HUD technology for work instructions in higher-risk tasks on site.

7   Connect BIM models with layers of active quality/environment/H&S/security information.

8   Monitor usage to find heat maps and dead areas of the IMS. Use analytics to drive improvements.

9   Connect IMS to other stakeholder management systems.

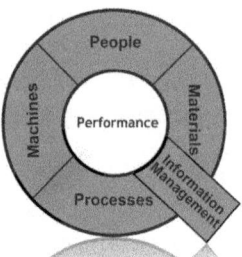

Notes

1 BSI, *BS EN ISO 9001:2015 Quality Management Systems: Requirements* (Milton Keynes: BSI Standards Limited, 2015), Section 1, p. 1.

2 BSI, *BS EN ISO 9000:2015 Quality Management Systems: Fundamentals and Vocabulary* (Milton Keynes: BSI Standards Limited, 2015).

3 Davis, W., 'SIPOC management: you're in charge. Now what?' *Quality Digest*, 20 August 2018. Retrieved from www.qualitydigest.com/inside/management-article/sipoc-management-you-re-charge-now-what-082018.html

4 BSI, *BS ISO 31000:2018 Risk Management: Guidelines* (Milton Keynes: BSI Standards Limited, 2018).

5 BSI, *BS EN ISO 10005–2018 Quality Management: Guidelines for Quality Plans* (Milton Keynes: BSI Standards Limited, 2018), Section 3.2, p. 2.

6 Akponeware, A.O. and Adamu, Z.A., 'Clash detection or clash avoidance? An investigation into coordination problems in 3D BIM', 21 August 2017. Retrieved from www.mdpi.com/2075-5309/7/3/75/pdf

7 McPartland, R., 'What is the Common Data Environment (CDE)?' *NBS*, 18 October 2016. Retrieved from www.thenbs.com/knowledge/what-is-the-common-data-environment-cde

12 Drones

It was Christmas 2002 and one of my son's presents was a tiny remote-controlled helicopter. It nestled in my palm and all the exciting packaging and joystick control suggested an exciting toy. It tended to spin around in circles and then crash. By lunchtime it was discarded. That was a typical response to flying remote-controlled 'toys': nothing more than a hobby. Even looking in wonderment at how professional hobbyists controlled little replica planes high over summer fields fitted into quirky nerd category.

I became a serious fan of 'unmanned aerial vehicles' (UAVs), commonly known as drones, when attending BIM Live 2015 where I tossed a business card into a competition goldfish bowl at a conference stall and a week later got the call that I was the new owner of a Parrot A.R. Drone 2.0! Fitted with cameras, 3-axis gyroscope, accelerometer and magnetometer. The stable video images as it took off, hovered, banked and turned, before flying over local houses and parks were impressive and reminiscent of movie shots taken from helicopters. The GPS and flight recorder meant that it was one of the best 'toy' drones available at the time and demonstrated to me future business capability. Several times I tried to explain to my line managers in construction businesses, how to use it, whether for security around perimeters or taking photographs and videos for construction inspections. To no avail.

Times change and these days, most large contractors will use drone subcontractors fitted with high resolution cameras to gather data on construction sites. From a hobby for a handful of enthusiasts, drone flying has become a serious business as the numbers of drones rapidly grew and the applications for using drones suddenly increased. The quality professional needs to understand the importance of the data streams and check that the data quality is being managed.

Aerial surveys by a drone in flight (Figure 12.1) can provide photos, video and a point cloud of data covering the designated site, which, in turn, allows photogrammetry (measurements) to be taken to show build progress against programme, to be visually assessed from the air. For a large site, that saves the time of walking or driving to the location and provides accurate measurements across surfaces.

However, even setting autonomous flight paths for the drone to follow, will still require drone pilots to maintain a line of sight with controls ready to take over in the event of an emergency.

Figure 12.1 Drone in flight.
Source: Ricardo Gomez Angel, flickr.com/photos/rigoan.

Privacy laws mean that a drone cannot be used to spy on members of the public sunbathing near a construction site. The images will end up being saved within a business cloud and all data needs to comply with data protection laws. Even inadvertent collation of video and images of people who have not explicitly provided consent may breach such laws and quality professionals need to be cognisant of such legal restrictions.

In the UK, a Permission for Commercial Operations (PfCO) issued by the Civil Aviation Authority (CAA), requires that there is a demonstration of 'piloting competence', sufficient understanding of aviation theory (airmanship, airspace, aviation law and good flying practice), passing a practical flight assessment (flight test) and an operational manual has been developed for conducting flights.[1] With specific models of drones, in addition, an Operating Safety Case (OSC) is required to demonstrate that the intended operation is appropriately safe. Pilot competence can be demonstrated by completing and passing courses run by approved National Qualified Entities (NQEs) who carry out assessments.

With the number of drone incidents steadily increasing year on year from twenty-nine in 2015, to seventy-one in 2016 and ninety-two in 2017, the UK government has woken up to the need to introduce safeguards against negligence and abuse.[2] In 2016, a UK study by the Department of Transport found that 'airliner windscreens could be critically damaged by mid-air collisions with 4 kg class quadcopter components, and 3.5 kg class fixed-wing drones with exposed metallic components at high, but realistic speeds'.[3] Tougher regulations, introduced in 2018 in the UK,[4] affirmed in legislation the existing guidance that restricts drones from flying above 400 feet and within one kilometre of airport boundaries. Any pilots of drones caught acting recklessly or negligently in a manner likely to endanger an aircraft or any person in an aircraft, will face imprisonment of up to five years. An online safety test will also be mandatory for drone pilots from November 2019, with fines of up to £1,000 for those who fail to comply.

In the USA, the Federal Aviation Administration (FAA) has created a drone registration support site, which states that:

> The FAA requires all drone owners to register each drone that is purchased weighing between 0.55 lbs to 55 lbs. If you meet the criteria to register an unmanned aircraft and do not register, you will be subject to civil and criminal penalties defined in the U.S. Government drone regulation terms.[5]

A remote pilot certificate is valid for two years and requires identification checks and passing an aeronautical knowledge test at a FAA-approved knowledge testing centre. The knowledge test covers regulations relating to small unmanned aircraft systems, loading and performance, flight restrictions, radio communications, inspection and emergency procedures, airport operations and health and safety.

However, some large businesses only take such certificates as the starting point, with additional training. Rio Tinto require their drone pilots to undertake flight training to understand risk assessments, flight procedures and technical manuals. Then they undergo simulator training before supervised site lessons to acquire at least twenty hours of flight time. Finally, they will sit a final exam before they are approved as a drone pilot.

The risk of accidents will rise with the use of drones and it is important to understand the need for professional training, not just to pilot the drone but also to ensure that operational procedures are followed for safety checks and knowing what to do in the event of an emergency. In Japan, in 2017, six people, including children, suffered minor injuries when a 4 kg drone crashed at an event in Ogaki, Gifu Prefecture.[6] The message is simple: undertake risk assessments prior to use and treat drones with respect, as any other piece of equipment or tool and follow the agreed operational procedures to minimise the risk of injuring people, damaging property and costly damage to the drone itself.

DroneDeploy is a software platform specifically aimed at drone flying and data capture for a number of industries. It is an ideal starter software for small and medium enterprises wanting to understand and develop drone use in engineering and construction. Images taken by the drone are uploaded to the software and a variety of construction-related activities can be carried out. A 2D aerial photograph of earthworks can be outlined and the software will then calculate the volume within a few seconds, rather than a survey team troop out on site, take the measurements and then return to the site office to make the calculation. Similarly, civil engineering companies have used DroneDeploy to carry out aerial surveys of existing highways to determine locations of deterioration, to plan out construction maintenance schedules.

In one case, a twenty-six-mile stretch of road was surveyed by Bolton & Menk for the city of Elko New Market in Minnesota in two days versus one week using vehicles and on foot.[7] The aerial survey captured much more granularity over traditional inspections with greater accuracy. Cracks in the asphalt were automatically plotted on maps, allowing calculations for new materials estimates for time required.

ReconnTECH has used DroneDeploy for efficient recording of underground utilities. The client employed ReconnTECH to survey an area around each telephone pole to establish the location of buried utilities so that new devices can be attached to the poles, creating Wi-Fi hotspots. Using a radio detection scanner, the underground cables are mapped on the pavement surface using paint markings. These markings are then surveyed and the GPS co-ordinates are manually plotted on a satellite photograph.

While the radio detection is still used with surface paint markings, the drone can then record the aerial images and turn the paint marks into cable routes, reducing each pole survey from two hours to forty-five minutes. For a survey of 424 poles across California, making 50 per cent efficiency gains, this is a huge improvement.

In 2016, McCarthy Building Companies invested in twenty drones after testing out the DroneDeploy software and created their Drone Champion Program to systematically begin integration into construction projects. The Program committee of representatives from across the business meticulously planned out over four months the development of safety protocols, operational procedures and a staff training course. The team consulted in-house risk management, insurance and legal experts to assess all aspects of implementing the technology. Once the green light was given, then a standard approach was developed to optimise results.

Each construction project is overflown three times per week recording video, with each flight taking around twenty minutes. DroneDeploy apps allow measurements to be taken on construction progress with data exported to Autodesk models. However, McCarthy still print off drone aerial images for display in site cabins to provide a clear progress update. Site staff and subcontractors also use the photos to identify poor housekeeping and where materials need to be tidied up. The point of collating the data is to facilitate greater collaboration and understanding that allow smarter decision-making.

For more complex construction projects requiring greater detail on images, then Pix4D is an option.[8] Drone flybys can analyse as-built site data with BIM models by creating point clouds from multiple angles and detect defects in construction structures. Orthophotographs can be taken (true, right angle aerial images with the same accuracy used in maps, rather than the slightly stretched versions that are taken as standard by drones as they fly overhead) to give better accuracy of construction sites. Digital model slices and 2D drawings can be overlain on the photos for comparison of scheduled work vs as-built, allowing tags to be added for queries and snagging.

Drone image data quality will be affected by the height it operates, the lookdown angle of the camera and the speed of the drone. Hence, it may be possible to fly a drone to capture the whole of a roof in a single photograph but will that image provide the level of detail need for the specified requirement of the data? It is also better to use software that can create orthomosaics rather than photo stitching techniques. Orthomosaics correct the camera perspective and the distance from the ground, as it flies overhead to give a true image. Sometimes photo stitching is painful when it shows the same image sliced through the middle but the edges not truly matching up.

Drones have been used in confined spaces and tunnels, such as Crossrail, to carry out inspections of shafts up to 50 m in height, avoiding the need for a mobile elevating work platform or scaffolding to access the shaft walls.[9] Thermal imaging by drones can be used for heat loss and water ingress in buildings. Infrared energy is part of the electromagnetic spectrum which is all around us in the environment and is mostly invisible to the naked eye with a small amount of light energy that we can see. Light is reflected off a surface but thermal energy can both reflect off a surface and be emitted. Emissivity is the level of thermal energy given off by a body. Humans, animals, trees and concrete have high emissivity and give off the thermal energy efficiently and hence are shown as bright colours in thermal imaging. Metals, however, act differently with low emissivity, although it can be effected by corrosion or a painted surface. The temperatures of two materials in close proximity may be the same but the emissivities may be radically different depending, upon the material and surface. Thermal imaging cannot see through glass or walls but since heat moves, the images can detect energy loss or changes.

Drone cameras need to be chosen according to their use and the mission requirements. Field of View (FOV) or size of ground coverage, magnification, and the detail of the image will vary and a balance needs to be struck when choosing the focal length lens. Bear in mind that images seen are usually relative to one another and not absolute. Atmospheric conditions, material types and surface conditions will affect the temperatures that may be shown.

While a rainbow colour palette looks the most dynamic and attractive, black, grey and white palettes can be more useful, in certain environmental conditions. In hot and humid conditions using White Hot or Black Hot colour palettes will aid the detection of subtle heat changes. Rainbow colour palettes (of which there are several versions, depending upon use) can provide useful imagery of targets where, for example, there is a high contrast between close proximity objects, such as electrical installations.

Equipment is designed to operate up to a maximum temperature. Drone thermal imaging can use isotherms, which will set an alert or flag to warn that the temperature for that specific equipment has been breached. That is useful when flying over a site where critical equipment is being used.

The drone roof surveys can identify any snagging visually but also any heat losses due to incorrect installation. Such surveys, though, will be impacted by the time of day and the weather, so planning is essential to access opportunities to record data. Also baseline roof surveys can be undertaken to set benchmarks for future comparative changes. For new roofs, this is a great way to understand how a specific roof that has recently been constructed is thermally acting.

Conversely, drones are also being produced in a variety of sizes, including micro drones. RoboFly weighs just one gram and is slightly larger than a house fly but faces a series of tough design and manufacturing challenges.[10] It can achieve wireless flight by a laser beam powering its tiny photovoltaic cell and its circuits convert the meagre seven volts into 240 volts. The battery life, however, is less than thirty minutes and its flight controls are currently very basic in flapping its wings for take-off and landing. But the engineers at the University of Washington who have created RoboFly are confident of building additional functionality. It is the first time that a sci-fi robot fly has been created that is wireless, and it opens up the possibilities of swarms of tiny drones, flying over a construction site with a diverse allocation of tasks from searching out leaks in newly installed gas pipes and water leaks in roofs through to material stock-taking and air quality testing in confined spaces. The potential is huge for inspection and testing anything and everything across a site, using tiny drones that will be relatively unobtrusive around operatives; although they may face the same fate as real flies, if they annoyingly get too close.

Some research suggests that there could be on average 76,000 drones air-borne across the UK by 2030, with a net impact of £42 billion to GDP, of which £8.6 billion is due to the construction industry.[11] With the functionality from a diverse set of drones likely to become available, inspection, testing and verification of work completed to standard could be undertaken by drones. Why send a quality engineer, to put on their PPE, check that they have a working tablet and walk out onto site past umpteen hazards, when a drone can more quickly take off collect digital data and return? It suggests that site quality professionals urgently need to understand both drones as major assets and at the same time that they may be taking away some parts of their job description.

With DroneDeploy reporting a 239 per cent increase in using drones on construction sites, compared to other sectors in their 2018 survey across 180 countries,[12] then it is clear that they are fast becoming more commonplace on projects large and small. With 10 per cent of the primary use of drones in construction being used for quality control, then it shows the need for quality professionals to get to grips with this technology is crucial, if they are not going to be left behind. The message is we need to adapt fast to be the guardians of drones, responsible for collecting and processing the data so that quality professionals enhance their digital capabilities.

Digital learning points

Box 12.1 Digital learning points: drones

1 Qualify as a drone pilot.
2 Be clear: what is the image data to be used for? Is there a clear process to quality assure the data?
3 Use drones for quality control inspections on site. Package up results with narrative report and drone data, including laser scans and video and photographic images.
4 Assess and track the transfer points of the drone data to assure the quality. Does the manufacturer's user manual provide recommendations on data management?
5 On quality audits check that:
  • the drone user manual is version-controlled and maintained;
  • drone batteries are at an optimum;
  • drone servicing is carried out.
6 Assess data security for any thumb drives or wi-fi used to download data.
7 Do drone dashboards show and record telemetries? Can this data provide improvements for future use?

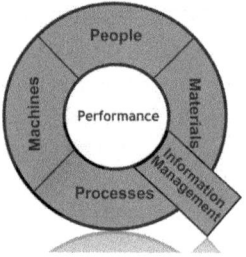

Notes

1 CAA, *Permissions and Exemptions for Commercial Work Involving Small Drones* (2015). Retrieved from www.caa.co.uk/Commercial-industry/Aircraft/Unmanned-aircraft/Small-drones/Permissions-and-exemptions-for-commercial-work-involving-small-drones/
2 *The Daily Telegraph*, 'Drone near-misses triple in two years', 19 March 2018. Retrieved from www.telegraph.co.uk/news/2018/03/19/drone-near-misses-triple-two-years/
3 MAA, BAPLA, DoT, *Small Remotely Piloted Aircraft Systems (Drones) Mid-Air Collision Study* (2016). Retrieved from https://assets.publishing.service.gov.uk/government/uploads/system/uploads/attachment_data/file/628092/small-remotely-piloted-aircraft-systems-drones-mid-air-collision-study.pdf
4 DoT/CAA, 'New drone laws bring added protection for passengers'. 30 May 2018. Retrieved from www.gov.uk/government/news/new-drone-laws-bring-added-protection-for-passengers

5 Federal Aviation Administration (FAA), Dronezone, *Welcome to the FAA Drone-Zone*. Retrieved from https://faadronezone.faa.gov/#/

6 *The Japan Times*, 'Candy-carrying drone crashes into crowd, injuring six in Gifu', 6 November 2017. Retrieved from www.japantimes.co.jp/news/2017/11/05/national/candy-carrying-drone-crashes-crowd-injuring-six-gifu/#.W2lpDy2ZOu4

7 DroneDeploy, 'Drones raise the bar for roadway pavement inspection', Blog, 2 August 2018. Retrieved from https://blog.dronedeploy.com/drones-raise-the-bar-for-roadway-pavement-inspection-9c0079465772

8 Pix4D, 'Measure from images'. Retrieved from www.pix4d.com

9 IW, 'Trial uses drone to carry out Crossrail shaft inspections'. 21 November 2016. Retrieved from www.infoworks.laingorourke.com/innovation/2016/october-to-december/trial-uses-drone-to-carry-out-crossrail-shaft-inspections.aspx

10 Booth, B., 'Slightly heavier than a toothpick, the first wireless insect-size robot takes flight', *CNBC News*, 3 November 2018. Retrieved from www.cnbc.com/2018/11/02/about-the-weight-of-a-toothpick-first-wireless-robo-insect-takes-off.html

11 PriceWaterhouseCoopers, 'Skies without limits'. Retrieved from www.pwc.co.uk//intelligent-digital/drones/Drones-impact-on-the-UK-economy-FINAL.pdf

12 DroneDeploy, *2018 Commercial Drone Industry Trends Report* (2018). Retrieved from www.dronedeploy.com/resources/ebooks/2018-commercial-drone-industry-trends-report/

13 Construction plant
Autonomous vehicles and telemetry

Autonomous driverless cars, taxis, trucks and buses have been making rapid progress in the USA, Japan, the UK and other countries. The UK Chancellor Philip Hammond stated in 2017 that the government's objective was to have 'fully driverless cars' without a safety operator, commercially on the roads by 2021.[1] The government believes the industry may be worth £28 billion to the economy by 2035, supporting 27,000 jobs.

The myriad of terms around 'driverless' vehicles can cause some confusion. *Driverless* is a non-technical vague term that seems to be applied to different driving technologies. *Automated* suggests controlled by a machine and usually is applied when there are technological aids on or around roads to control the vehicle. *Co-operative* technologies have also been used where there are communications to the vehicle to aid control. *Autonomous* is used when there are no other road aids and the vehicles are independent. The US-based Society of Automotive Engineers (SAE) has set out different levels of driverless technology:[2]

- Level 0 – no automation with some 'enhancements by warning or intervention systems'.
- Level 1 – driver assistance for certain functions with the steering and pedals, e.g. cruise control.
- Level 2 – partial automation with additional features, such as 'hands off' driving, e.g. Tesla's autopilot.
- Level 3 – conditional automation – under certain conditions the car can take over driving functions that will allow 'eyes off the road', such as changing lanes.
- Level 4 – high-level automation with 'hands, off, eyes off and sometimes mind off'. Google's self-driving car was at this level, in late 2017. However, a driver needs to be sitting in the driving seat and available to take over, if required, under, say, extreme weather conditions.
- Level 5 – no human driver required under all road driving conditions.

For most car manufacturers, when they talk of driverless vehicles, they usually refer to Levels 3 or 4. Honda and Toyota have targeted 2020 for self-driving on the highway with Volvo a year later. Fiat-Chrysler have a goal of 'some' self-driving cars by 2021 and Ford have been talking about 2021 for delivering 'true self-driving'. GM have been cautious of setting out a hard timeline but have said that they are aiming to be the 'first high-volume auto manufacturer to build fully autonomous vehicles', suggesting some time after 2019. Tesla cars in 2017 already have enhanced autopilot with 'full self-driving capability', in 2018. Elon Musk has been saying for two years that a hands-off drive between Los Angeles to New York City would be forthcoming, without it happening yet.

Whatever the final production runs and proven status of such Level 3 and Level 4 cars, it appears that unless conditions change, then most major manufacturers will have some types of autonomous driving cars on the roads in the USA, Europe and Japan, by the early 2020s. After decades of prototypes and promises of driverless cars, the rubber is about to hit the road, in a big way.

'We predict that construction sites will be unmanned in the future,' said Chikashi Shike, president of Komatsu's Smart Construction Division in 2018.[3] A bold claim and given the level of hype over the years for digital construction, perhaps should be treated with some caution as to which year he means by 'future'.

For the construction industry, autonomous vehicles can deliver important savings by using sensors to run more efficiently through monitoring the performance of the vehicle; tyres lasting up to 50 per cent longer and fuel efficiency increases. Given knowledge of the terrain route, the autonomous vehicle can plan an optimal driving technique, micro-managing each second of the journey for optimum effect. They can continue to operate past times that would require manual operators to take a break, meaning that they can operate for longer times and overall take less time to complete a task. On a site, autonomous vehicles face different challenges from typical road conditions and need adapted technology to solve them.

Komatsu has been blazing a trail with autonomous vehicle technology that is proven in the field. Starting in 2015, they partnered with Skycatch[4] to develop Explore1 drones that fly autonomously over construction sites, to create highly accurate 3D maps that feed into digital models. Take-off, flight routes and landing are all automated, freeing up time for operators. The data can then be used to direct Komatsu's robotic earth-moving equipment.

Komatsu's intelligent excavator[5] range can deliver more than 30 per cent increase in productivity[6] as the machine digs with precision to the required levels and moves earth according to optimal algorithms, such as avoiding over-digs in trenches. They reduce the demand for a wide variety of machines on an earth-moving contract as, for example, the accuracy of an intelligent excavator will then only require a small crawler dozer to clear up the remaining 'crumbs' to the correct grade, instead of needing another sweep by a larger excavator or dozer. Crucially, Komatsu spend a lot of time and effort on customer support to help them optimise the use of the intelligent machines to achieve quality performance. Classroom and field training help get the operators to a high capability, as soon as possible.

Do robotic excavators make drivers redundant? Sometimes, yes, but those same individuals can learn to become an operator of an autonomous drone and an autonomous dumper truck and add significantly greater value and productivity to that business, surveying the land and then shifting more dirt, at less cost and in a quicker and safer way. Their knowledge remains essential but the way they use that knowledge will change.

Rio Tinto's 'Mine of the Future' programme is a demonstration of successful autonomous vehicles being used on site to move ore and waste material.[7] Since it was first trialled in 2008, the autonomous haulage system (AHS) has been a great success with over one billion tonnes moved autonomously. One quarter of the trucks used at Pilbara in Australia are now autonomous vehicles, with numbers rising to 140 trucks planned to be operating by 2019.

A central operations headquarters in Perth can control huge Komatsu dump trucks 1,500 km away in Pilbara, Western Australia. Wireless gaming controllers are used to control rock-breaking machines and one operator can manage a fleet of driverless trucks. The human capability needed leans on skills that gamers find intuitive and frankly more fun than traditional computer applications. It encourages younger entrants into the construction industry and, together with more interesting and stimulating tasks, improves the chances of retaining talent within the industry.

A construction site can be a physically demanding place. Driving off-road is particularly wearing on the driver's neck, back and arms, even with modern power steering. By sitting in a comfortable seat off-site and 'driving' (it could be in a site office a few hundred metres away or on the other side of the globe), improves the quality of work life for the operator and their safety, by removing their physical presence from a high risk environment. Productivity increases as less time is spent on the operator moving to and from the vehicle and the overall process is more efficient.

Volvo Construction Equipment have developed an LX1 prototype electric hybrid wheel loader which has achieved hundreds of hours of real-world testing, proving a 50 per cent improvement[8] in fuel efficiency compared to its conventional front-end loaders and a reduction in noise pollution for operators. The LX1 can do the work of a larger wheel loader due to its more efficient design. At a $22 million research project at a quarry in Sweden in partnership with Skanska, the Swedish Energy Agency (SEA) and two Swedish universities, Linköping University and Mälardalen University, it is estimated that by using machines like the LX1, energy usage could be reduced by 71 per cent. With Caterpillar and Hitachi Construction Machinery also actively testing self-driving vehicles at mining operations in Australia and the start-up Built Robotics, created by Noah Ready-Campbell, formerly at Google, developing its autonomous track loader (ATL), called Mary-Anne, there is real progress in innovative robotic technology on construction sites.

According to a report by the Intergovernmental Panel on Climate Change (IPCC), the building sector was responsible for 32 per cent of the global energy consumption and contributed a quarter of the global total CO_2 emissions.[9] As part of strategic low carbon targets for the industry, the LX1 fuel efficiency figures could also play a key role in developing a more sustainable construction industry.

By combining electric technology with robotics, Volvo have developed the HX2, a prototype autonomous, battery-electric, cab-less load carrier, with a predicted 95 per cent reduction in carbon emissions and up to a 25 per cent reduction in total cost of ownership.[10] Looking like a headless dump truck, it parks alongside the LX1 to await earth or aggregate materials being loaded before moving off to its designated destination to transport the materials.

Volvo CE has also partnered with LEGO to develop a construction robot for children. The Volvo Concept Wheel Loader ZEUX has an 'eye' on the roof, which was designed following feedback from children.[11] When crossing a road, it was agreed that being able to make eye contact with the driver of a vehicle was important to establish a relationship and give a clear indication that the vehicle was slowing down to a stop. While there are less road crossings on a construction site, there certainly are many instances of human and vehicles interacting and unfortunately with the significant risk of an accident. That human-machine interaction in ZEUX was something that the toy maker wanted to capture and hence the 'eye' feature. While some may be sceptical in thinking that this is just a marketing ploy to sell toys and for Volvo improve their brand recognition with children, I believe it highlights the evolution of construction industry maturity.

The ZEUX loader will make it normal for both boys and girls to expect to see autonomous, electric vehicles in construction. It will engage them at an early age to think about construction at the cutting edge of technology and will help knock down male stereotyping to encourage more women to enter the industry and extend the pool of talent.

How the construction plant perform on site may have a direct impact on the performance of the work. There will obviously be time delays and risk to the project programme but also if construction plant do not perform as expected, they can affect quality of the work, whether it is excavating, material logistics or crane movements.

Telemetry is a form of communications whereby monitoring data is sent remotely to other equipment. Global positioning system (GPS) technology has been available for decades on construction vehicles to monitor location and enable satellite navigation. That provided some basic data on logistics, for example, tonnage per hour for quarrying trucks. As vehicle engines developed with computer management systems, so the type and volume of performance data increased, with information on service diagnostics and pollution emissions.

These days, real-time data is being streamed constantly from a range of construction plant vehicles, including excavators, lorry loaders, gantry, mobile wheeled and tower cranes, forklift trucks, road rollers, ready mix concrete trucks and earth-moving plant; dumpers, bulldozers, graders and scrapers. Other machinery and equipment include mobile compressed air units, bitumen mixing and laying plant, concrete mixing plant, hoists, conveyors and mobile elevating work platforms (MEWPs).

Environmental telemetry data covers fuel consumption, idle time vs work time and emissions, and safety data can improve adherence to driving standards and include proximity alarms for site operatives moving into high risk zones around construction plant. Financial data can be gleaned from the usage, downtime and efficiency of the plant and knowing when plant is returned to the yard from GPS can allow automated billing to clients.

Quality management data is the data that monitors impacts on the performance of the construction processes and creating the final built environment. Maintenance of the plant is one example of useful data that should be assessed by the quality professional as one element of the overall quality audit, in the same way that measuring instruments are checked for calibration. When construction excavators are removing a certain tonnage of material, the spoil heap volumes and bucket loads can be reconciled to improve the accuracy of the measurements. Likewise, excavated depth levels can be ascertained from construction plant telematics to compare with laser-scanned drone data.

Unlike earth-moving equipment, generators and compressors are not being constantly operated by people and, as such, understanding their performance and having prompt alerts are very useful to optimise their work time. Through telematics, quality performance data can be analysed for cause and effect. A generator going down at a critical moment on site can create defects and/or re-work. By monitoring remotely the generator's performance, such outages can be minimised and ensure runtime is maintained throughout a day's work before refuelling. If a pneumatic concrete vibrator while compacting concrete ceases to operate from the air compressor failing, the quality of the cured concrete may be affected.

Telematics from ready-made concrete mixing plant allows back-end system data to be integrated with information captured in the field that informs real-time decisions. Delivery tickets for ready mix concrete trucks that in the past used to be masses of tissue-thin, pink slips of paper that invariably either went missing or were splattered in wet concrete, obliterating bits of writing, have now become secure digital data, within fleet management platforms. Live streaming provides a wide variety of data with great accuracy that can both produce standard management reports and trigger alerts if predetermined limits are breached or fault codes detected.

Ready mix concrete (RMC) truck deliveries are tracked by GPS to optimise arrival times at site, even allowing for traffic hold-ups and driver breaks, which maintain concrete mix standards for scheduled pours. The data also allows interrogation that in the unlikely event of later concrete tests failing specifications, the timeline, from creation at the batching plant through delivery to the pour, can be digitally investigated. Such data should be in accordance with the Quality Scheme for Ready Mixed Concrete (QSRMC) Quality and Product Conformity Regulations, which incorporate the requirements of BS EN ISO 9001:2015 made applicable to ready mixed concrete and combining BS EN 206:2013, BS 8500-1:2015 and BS 8500-2:2015. The QSRMC regulations are an additional concrete-specific assurance that the product may be required to conform to, in the given contract specification.

With all these applications of construction plant telematics, the point is not to go chasing data for the sake of it but rather for the quality professional to be aware that such data may exist (or in particular cases they wish to specify at the beginning of the project) and it can be useful in root cause analysis (RCA) and driving continual improvement. Importantly, using Big Data can also improve predicting where there is greatest risk of quality management issues. The more intelligent data collated, and the better the analysis (from BI tools or AI), the easier it is to identify past and future trends.

Helping construction personnel to better understand the reasons behind substandard work or heightened risks to quality standards, provides an overall improvement to construction risk management that may also feed into health and safety, environmental, security, business continuity and other business assurance risks.

Digital learning points

Box 13.1 Digital learning points: construction plant: autonomous vehicles and telemetry

1   Assessing the capability of a driver for Level 2 autonomy, who may be based on the other side of the world in an office from where the vehicle is operating, will require demonstrable evidence other than a driving licence. Will a new 'gaming'-type driving qualification be required? But if the vehicle is a Level 3 or Level 4 and self-driving, then the capability of the AI becomes more appropriate than a human capability.

2   Understand what kind of telemetry data is available for each type of construction plant. Does any of the data report useful quality performance? If it does, then check that it is being monitored by the plant contractor and used.

3   When autonomous machines are self-driving, only the AI process in its very basic sense could be mapped. How will processes be assessed with an AI 'black box' of tasks between an input and output?

4   Does the telemetry have documented processes for data quality? How can the operator be sure that the telemetry measurements and information are accurate?

5   Are the machines being properly maintained? Many of the checks will focus on safety but some will report on serviceability. Downtime is waste. How are the plant contractors trained on the plant before operations?

6   When materials are being moved, is the plant protecting those materials? For example, lifting bricks or blocks by crane or hoist. Are construction materials being moved efficiently by the plant?

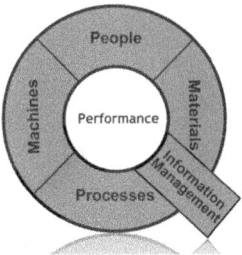

Notes

1 BBC News, 'Hammond: Driverless cars will be on UK roads by 2021'. 17 November 2017. Retrieved from www.bbc.co.uk/news/business-42040856
2 Dyble, J., 'Understanding SAE automated driving: Levels 0 to 5 explained'. *Gigabit*. 23 April 2018. Retrieved from www.gigabitmagazine.com/ai/understanding-sae-automated-driving-levels-0-5-explained
3 CIOB BIM+, 'Komatsu takes first step to the autonomous construction site'. 17 December 2017. Retrieved from www.bimplus.co.uk/news/komatsu-takes-first-step-autonomous-construction-s/
4 SkyCatch, 'All-in-one drone data solution for enterprise'. Retrieved from www.sky-catch.com
5 Komatsu, 'Komatsu intelligent machine control: the future today'. (2017). Retrieved from www.komatsu.eu/en/Komatsu-Intelligent-Machine-Control
6 Ibid.
7 Rio Tinto, 'Smarter technology'. (2018). Retrieved from www.riotinto.com/our commitment/smarter-technology-24275.aspx
8 Volvo CE, 'LX1 prototype hybrid wheel loader delivers 50% fuel efficiency improvement'. Press release, 7 December 2017. Retrieved from www.volvoce.com/global/en/news-and-events/news-and-press-releases/2017/lx1-prototype-hybrid-wheel-loader-delivers-50-percent-fuel-efficiency-improvement/
9 Lucon, O., Ürge-Vorsatz, D. *et al*. 'Buildings', in IPCC, *Climate Change 2014: Mitigation of Climate Change* (Cambridge, Cambridge University Press, 2014). Retrieved from www.ipcc.ch/pdf/assessment-report/ar5/wg3/ipcc_wg3_ar5_chapter9.pdf
10 Volvo CE, 'Volvo CE unveils the next generation of its Electric Load Carrier concept'. (2017). Retrieved from www.volvoce.com/united-states/en-us/about-us/news/2017/volvo-ce-unveils-the-next-generation-of-its-electric-load-carrier-concept/
11 Volvo CE and LEGO®, 'Volvo CE: Introducing ZEUX in collabortion with the LEGO® Group'. (2018). Retrieved from www.youtube.com/watch?time_continue=25&v=3uJCgt_2Y4o

14 Robotics, lasers and 3D printing

Robots

It sounds like a drill in short staccato bursts, as the robotic arm doggedly drops from its hoist slung across an expanse of latticed steel bars, pinches a cross-section and twirls the wire around before cutting, rising and repeating the task at the next intersection. Welcome to the Tybot,[1] that must have one of the most boring jobs in construction robotics. With the pain from musculoskeletal injuries affecting most construction operatives – 52,000 from a total of 80,000 self-reported injuries or illnesses per year – this is a process ready for automation to improve health and safety.[2] For small rebar tying tasks or areas that are more awkward to access, then using operatives to hand tie or use automatic tying guns is likely to remain the cheapest and easiest solution (although with continuing health risks). However, for large areas of rebar, then using a robotic solution frees operatives to undertake other tasks and reduces the probability of injury, provides a high confidence level of meeting quality standards and removes the need for manual quality control inspection.

Hadrian X,[3] Blueprint Robotics[4] and SAM100[5] are all robotic systems that can lay either bricks, concrete or pre-fabricated panels. The robotic systems replace large teams of operatives by more efficiently creating structures, to meet standards and specifications, with only minimal technician support and direction.

There is a history of robots being developed for construction. In 1904, John Thomson patented the first bricklaying device using pulleys and levers.[6] In 1967, Pathé News filmed the 'Motor Mason', which was heralded as the modern solution to more productive bricklaying.[7] It was a mechanical machine moving on rails with a hand-turned wheel laying mortar and lifting and rotating bricks into place in a wall. It worked efficiently on straight walls but could not turn corners or place bricks in patterns. However, SAM100 can lay up to 3,000 bricks a day but still needs help stocking the bricks in the machine and clearing up excess mortar.

In Switzerland, in 2015, the 'In-Situ Fabricator' created by ETH Zurich's National Centre of Competence in Research (NCCR) Digital Fabrication lays bricks in free-flowing curves using a wheeled robotic arm but so far does not apply mortar, restricting its practical application.[8] Nevertheless it has solved the free moving challenges of a robot and may well be technology that is used in future robots that need to move around a construction site, laying bricks. Two years later, the In-situ Fabricator had progressed to creating complex-shaped mesh reinforcement designs.[9]

The Brokk range of demolition robots uses remote control technology married to artificial intelligence that can efficiently calculate the optimum process to take down and remove a structure, saving time and money, with less risk to operatives.[10]

The HRP-5P humanoid robot, built by Advanced Industrial Science and Technology Institute (AIST), in Japan, can pick up and fix plasterboard (dry walling) to a timber frame.[11] It is not that this robot can necessarily outperform an experienced joiner but rather demonstrates that the technology is rapidly reaching a point where it will begin to offer a viable alternative.

Honda's ASIMO (Advanced Step in Innovative Mobility), shown in Figure 14.1, is the product of two decades of ingenuity and research by engineers to create a humanoid robot.[12] At 1.3 ms tall and weighing 50 kg, it looks like a small astronaut and it can recognise faces and walk around its environment unaided using machine vision. While ASIMO is designed for the domestic environment, particularly to aid Japan's ageing population, its human-like capabilities are impressive and suggest that such robots navigating the work

Figure 14.1 Honda's friendly humanoid robot, ASIMO.
Source: @franckinjapan.

environment are not too far away. After decades of film fantasies offering visions of robots everywhere, finally, we may just be on the cusp of making it a reality.

As technology improves and robots start to appear on construction sites, perhaps in the 2030s, producing them will lower their unit cost and the appearance of humanoid robots will become commonplace.

With all these types of robots, the quality professional needs to understand the design processes that underlie them. Why? How can we contribute to developing such systems, if we do not understand them? How can we agree that they are quality assured, if we have no idea of the underlying algorithms and software? Quality professionals do not necessarily need to understand at a deep level the programming, but they do need to appreciate fundamental quality management of algorithms lest they do not perform as expected.

If the robot has replaced a human to deliver a particular task (e.g. laying bricks), then the quality assurance check is similar to what would be asked of a human: competency. Can the robot (and hence the software) demonstrate that it has the 'competency' to deliver the required performance before laying a single brick?

A qualified bricklayer may have a Level 2 Certificate or a Level 3 Diploma in bricklaying and an appropriate Construction Skills Certification Scheme (CSCS) card. Therefore, if a contractor proposes to automate bricklaying for a project, then as part of the tender documents, it should include a conformance certificate from Hadrian X's manufacturer, FBR or, the contractor needs to attest to its performance. Regardless, the project management team should understand that it needs to proactively check that such machines can deliver the required performance before they start work.

Doxel[13] software can be linked to an autonomous tracked robot the size of a shoe box that roams and navigates through a construction site each day (even climbing steps) while scanning the as-built environment to compare progress against program in a digital model. The robot is equipped with high definition cameras and 'light imaging, detection and ranging' (LIDAR) sensors, similar to driverless vehicles. Deep learning algorithms can instantly calculate the percentage of visually accessible structural work, electrical, mechanical and plumbing works completed.

Doxel can also identify work that may be out of tolerance and conduct visual quality control inspections with much greater reliability and accuracy. The ongoing assessments and inspections provide site managers with updates on progress and allow them to manage resources more effectively, while reducing costs and improving quality. Such updates can increase the lead time of inspections by a factor of six, giving site management more warning (and thinking time) of issues and problems.

On case studies, it has proven to have a cost-at-completion of 96 per cent accuracy, 38 per cent increase in manual productivity and be 11 per cent under budget.[14] Even if these are maximum improvements, they demonstrate it is worthwhile investigating and piloting Doxel on site. Challenges will include timing of assessments and inspections to minimise the risk of people being in the vicinity and ensuring that housekeeping is to a high standard to prevent the robot hitting obstacles.

Laser scanning

Until recently, the best surveying equipment would have been the Total Station that integrated an electronic theodolite, called a transit, with an electronic distance measurement (EDM) system. The Total Station, while being very accurate, is relatively slow, capturing one data point at a time. Such additional time is waste in the era of Digital Quality Management.

High definition surveying or laser scanning is the process of firing multiple lasers that accurately and rapidly capture points in space on physical structures that can be used to create a virtual model of the real world in a 'point cloud'. The market for laser scanning is estimated to be growing steadily, with the global construction lasers market estimated to have reached $2.4 billion in 2017, and is estimated to reach $3.3 billion by 2025.[15]

For renovation projects where the only records may be redlined paper drawings with limited use, a laser scanner can rapidly record the interior and exterior of the building to assemble a new BIM model as a baseline for future building improvements. For new build, the laser scanning can check as-built construction versus digital model design to provide highly accurate inspections with real-time information that will either confirm acceptance of the as-built or identify non-conformances. With re-work typically costing up to 12 per cent[16] of a construction project, laser scanning has demonstrated it can reduce such costs to 1 per cent.[17]

The laser rotates horizontally and mirrors bounce out the lasers at ninety degrees. The machine will slowly spin around its y axis, so that around one million points per second are scanned out in a sphere shape to an accuracy of +/–1 mm with ranges of up to 350 m. The batteries can last for half a day of surveying before recharge and a touch screen allows intuitive commands. HD cameras often are built into the scanner to allow 160 megapixel resolution images to be captured that can overlay natural light to the points cloud, providing even greater overall imagery. The scanner is usually fitted with a range of sensors, inclinometer, GPS, compass and altimeter that compensate and refine the data for improved accuracy and reliability.

Small reference spheres or orbs are placed around the site to optimise the scan efficiency. Typically grapefruit-sized, plastic balls with magnets are placed on steel beams or fitted with small pedestals to stand on floors. These reference spheres remain in place whenever the laser scanner moves to a different vantage point, to act as a transformation link between the scans, which are stitched together by a registration process. To be less obtrusive on existing buildings where people may be working or hospitals with patients, ping-pong balls have been used! Chequerboards also provide the same functionality as reference spheres. These are black and white alternate squares on a board that are temporarily attached to walls or ceilings.

All this sounds simple but on sites with moving people, vehicles, animals and the vagaries of the weather, achieving high accuracy to meet client and contract requirements can still require experience, expertise and lots of patience.

The data is imported from the scanner into software that can then process, register, view and finally export the data in different file types to other CAD programmes. Processing allows filtration to remove (although no data is ever lost) very dark images, such as fine cracks in walls, clean up edges of data and remove stray points, such as dust in the air, which distract from the point cloud visualisation. At all stages of processing, the data can be saved. If there are gaps or anomalies between scans and the software has not been able to overcome them using the targets, the data can be manually manipulated to piece together the same objects using the trained eye.

Registration will then pull together the scans through many iterations as it identifies the same points in space from different scans. Viewing can create certain perspective views for presentations or meetings to save time from running through a complete scan to find a certain perspective. Exporting may be a challenge with the size of files and designers may slice up the models to send and have it 'lofted' back together again into one model inside a different program. Or clipped boxes of 3D information may be sent of a particular part of the model. Different options provide for different collaborative tasks and reduce the time needed in sending large amounts of data repeatedly.

For different site environments, different types of scanners may be used from static mounted to mobile, hand-held scanners. The resultant point clouds can still all be imported and registered together to provide one holistic imagery. On large construction sites, the file sizes can become seriously large, which may require minutes or even hours to process.

Reports can be generated to show the accuracy of the data capture and registration process. These can be colour-coded according to pre-chosen thresholds to instantly highlight issues. Even if the maximum error is, say, 5 mm and the overall accuracy is averagely reported at 2 or 3 mm for a site covering the size of a football pitch, it may not mean though that each feature is within agreed tolerances. A lamp post, for example, up close may show as two separate posts instead of one or a line of kerb stones may be misaligned between scans. In this way of manually diving into the data visualisations, an experienced user can assess the quality. By running scan management, quality control can be undertaken to re-run the registration from cloud to cloud with the software removing the misalignments itself and tightening up the image.

Typically, this may be where there is no systematic recorded quality control results. It demonstrates the importance of quality professionals understanding digital information management processes and setting or advising requirements and quality thresholds. If a BIM Manager appreciates quality management, then they will keep a folder of quality control reports and will be able to confidently describe to a quality auditor the process of registration and how the maximum and mean accuracies were tightened up to within a specified tolerance of a digital laser-scanned model. If such records are not maintained, then the quality auditor is left asking for subjective information on the quality control process. That is always open to interpretation and bias and may not give an accurate description of the time taken nor demonstrate mistakes and opportunities for improvement. By maintaining the records, a timeline can be established and the numbers of changes made, which can be valuable information on, for example, how the registration process can be improved. That can be used to feed back into the training of BIM professionals and software manufacturers. It also helps reinforce the culture of learning and quality management within the design team, all helping to drive continual improvement each day.

Drone mapping is a development that takes the laser scanning technology into the sky and allows fly-overs to map a site. Static photographs can be useful but limiting, and the drone provides a full sweep without the need to flick back and forth between photos. The scanning can provide quantities of materials, such as earthworks, saving time and money on sending out a surveying team. Software such as DroneDeploy allows filters to remove shadows on the ground and ensures accurate outlines are established, before making volumetric calculations. The process is simple and straightforward and very quick. The DroneDeploy app has a security and safety feature that it will not work within a designated No Fly Zone, mitigating the risk of incidents.

3D and 4D printers

In 2004, Professor Behrokh Khoshnevis, of the University of South Carolina, developed the first prototype 3D printer, called contour crafting, to print a wall. After many experimental studies in various laboratories, the 3D printer technology was proved by printing out buildings using concrete mixes around the globe.

A 'Research and Design by Doing' project created the 3D Print Canal House in 2014 with an international team of partners beginning to print a canal house in Amsterdam.[18] Rooms are partially printed onsite and then fitted together before rooms are interconnected to create a floor. In the same year, the Chinese company Winsun announced it had 3D-printed ten concrete houses in twenty-four hours at just £3,200 each, using a printer that was 32 m long, 10 m wide and 6.6 m tall. A year later, Winsun were the contractors for the 'Office of the Future' in Dubai by printing in seventeen days a small 250 m² building. The services and interiors took a further three months but it claimed labour costs for a comparative building were cut by half.[19]

Dubai has set out an ambitious, radical agenda for 3D printing of city buildings with His Highness Sheikh Mohammed bin Rashid Al Maktoum, Vice President and Prime Minister of the United Arab Emirates (UAE) and Ruler of Dubai, launching the 'Dubai 3D Printing Strategy' in 2018. The strategy aims to make the UAE the leader in the technology with each new building in Dubai to be 25 per cent 3D-printed by 2025.[20]

The claims for the first habitable 3D constructed house are made in different countries. In 2018, the Van Wijnen construction company in collaboration with the Eindhoven University of Technology, announced it would be printing homes with three floors and three bedrooms near the city of Eindhoven.[21] Only the exterior and inner walls will be printed off-site and then fitted together on plots near to the city's airport at Meerhoven.

In Nantes, France, another experimental 3D four-bedroomed house was created in 54 hours by the city council, the University of Nantes and a housing association,[22] using Batiprint3D[23] technology. The cost of £176,000 was claimed to be 20 per cent cheaper than using conventional materials and processes. A family duly moved into the property. The polymer-based material demonstrated the capabilities of this technology by printing curved walls that were both aesthetically pleasing and wrapped around 100-year-old trees so that they were not damaged. Two layers of expansive sealing foam are printed with a gap in between where the concrete material is printed over a minimal reinforcing bar, using the foam as a formwork. The foam, however, remains in place to provide excellent thermal insulation.

The Centre for Additive Manufacturing (AM.NUS) at the National University of Singapore has created pilot studies for printing simple toilets in India, given the urgent hygiene improvements that are needed in some communities.[24] Results indicate that the construction time is halved to five hours and the costs are reduced by 25 per cent.

ICON,[25] a start-up in Austin, Texas, unveiled a printed home in March 2018 of 600–800 sq. ft home constructed in under twenty-four hours for less than $4,000.[26] They are working with a not-for-profit, New Story, to find housing solutions, for disadvantaged communities in Bolivia, Haiti and El Salvador.

The DFAB House project by the National Centre of Competence in Research (NCCR) in Switzerland 3D sand-printed an 80 m² lightweight concrete slab in 2018, which is half as heavy as an equivalent conventional concrete slab.[27] At its thinnest point, it is just 20 mm, demonstrating the precision and functionality of the pre-fabricated technology. The project is experimenting with a range of digital fabrication technologies.

After making an announcement in 2015 to print the world's first bridge, it was duly unveiled offsite in 2018 by MX3D[28] and a consortium led by ARUP, Lloyds Register, The Alan Turing Institute, Autodesk, Faro and others. The stainless steel, pedestrian bridge[29] will cross the Oudezijds Achterburgwal in Amsterdam. It took many months of design and re-design in order to meet the exacting requirements of safety, the challenges of the medieval canal walls and the complex geometry of the 12 m span. A network of sensors will be installed in the bridge to provide a stream of data on the structure's displacement, vibration and strain from people crossing it, together with environmental information on air quality and the impact of the weather to monitor the bridge's health over time. This data will then inform the designers how the bridge is working in situ for future feedback to the bridge itself, so it 'understands' its own environment. In turn, by comparing the physical bridge's performance against its digital twin, learning can be gleaned for future 3D bridge designs.

4D printing[30] takes the 3D printer technology and tweaks it, so that the materials can adapt to changes over time: heat, vibration, sound or moisture. These changes can be made to physical and biological materials to alter the shape and compute from silicon-based matter. Self-assembly processes build ordered structures only through local interaction. Practically, this could mean that underground pipework could change shape to adapt to capacity or flow rate, according to demand.

While the first 3D construction printers are available now (see Figure 14.2), the technology continues to be refined and questions on the environmental friendliness of some of the materials need to be satisfactorily answered. Large buildings and structures with different designs are not yet being printed nor are on-demand components on a construction site but in the space of a few years great progress has been made and the quality professional needs to get ahead of the curve.

Balfour Beatty has been using 3D printing to create scale models of components to show clients, which are more effective in presentations than the average PowerPoint on a 2D screen.[31] An 'art of the possible' called Project AME, 3D printed the carbon fibre cab, steel arm and aluminium heat exchanger of an excavator to demonstrate what can be printed.[32] A surprising and challenging 'knitted concrete' sculpture in Mexico has demonstrated that there can be all kinds of weird and wonderful techniques and materials appearing in the built environment in the future. Created by ETH Zürich, a knitting machine produced a textile-like, shell design that is used as part of the formwork, for a concrete layer to be added.[33]

As these printers increase in use and applications, quality professionals may find changes in quality control. Will concrete slump and cube tests become

Figure 14.2 3D printing of building design components will become commonplace in future.

Source: zmorph3d.com.

obsolete? How will polymer materials be satisfactorily tested and certified for strength? Will sensors be embedded as standard in the printed materials that can instantly demonstrate compliance with building codes and manufacturing technical data sheet standards, and learn from its surroundings? Until these questions can be answered through automated quality control, as building components or whole buildings are printed, the quality professional needs to apply existing building codes and quality management best practices to 3D and 4D printing processes.

The project quality plan (PQP) should set out the role of the competent person who is responsible for supervising the setting-up and use of the 3D and 4D printers. The concrete mix must be specified to meet building regulations and contract-specific requirements. The concrete mix should be inspected and tested as usual for workability, strength and plastic density. The method statement should be written by those technically capable of using the robot, pump, hoppers and the extruded concrete material.

Cement and aggregate hoppers hold the materials separately before they are carefully dry mixed using screw conveyors with admixtures being added, if required. The mix has water added as it passes through a second screw conveyor before entering the pump, which pushes the mix into the robot. The nozzle end of the robot controls the extrusion, forming it into the desired shape layer by layer, until the specified height is reached. This gives superb, on-site precision with little waste as it is printed only where it is designed to go. Not only can walls be 'piped out' like giant icing nozzles in concrete but individual structural components can be printed and then fitted together like Lego bricks.

The challenges with the design mix are getting the balance right between pumpability and build-ability, and online monitoring is needed to ensure the pumped material meets the required product standards.

The bubbly finish of the 3D concrete mix may not always appeal aesthetically to customers. But the low cost, good thermal insulation and rapid build should entice enough customers to be profitable for companies. As the technology matures, then the final finish will improve and the size of 3D-printed buildings will grow.

Taking the automotive sector as an example, the first prototype spot-welding robots entered service in GM factories in 1961. By the late 1960s and early 1970s, robots were being developed by universities to be more versatile and hence more useful to automotive manufacturers. By the 1980s, the big three car manufacturers in the USA were investing billions of dollars in robots on assembly lines. If construction follows the same trajectory, then robots, including 3D printers, could start to become a common sight on large housebuilding sites by the early 2020s and large infrastructure sites by the mid- to late 2020s. SmarTech has predicted a construction 3D printing market rising from just $70 million in 2017 to $40 billion by 2027 comprising of $150 million for materials, $3.5 billion for machinery and $36 billion in services and applications.[34]

As with any equipment that affects quality of the final construction, so the 3D printer needs careful calibration in the three axes: X, Y and Z, and the extrusion must be within agreed manufacturer's tolerances and the material should be printed in accordance with the design specification. If the concrete wall is printed a few centimetres in the wrong place, it will cause the same quality problems as concrete that is poured into formwork that has been set out inaccurately. The quality professional should ask to see such 3D or 4D printer calibration records and certificate in good time as part of the due diligence, before the printer commences printing. Inspection and testing should be based on the digital records generated that there is evidence that the material has been correctly printed to the agreed measurements.

Digital learning points

Box 14.1 Digital learning points: robotics, lasers and 3D printing

1 Understand how robots, laser scanners and 3D printers are operated. Does any of the performance data report on quality performance? If it does, then check that it is being monitored by the robot or 3D printer contractor and used to improve performance.

2 Does the use of the robots, lasers and 3D printers have documented processes? Are user manuals available to be referenced to documented procedures? Are there links to current user manuals? How can the operator be sure that the robots and 3D printers will deliver to the required quality performance?

3 How is registration of laser scans quality managed? Is there an audit trail of changes to the digital point cloud models through saved scans?

4 Are the robots, laser scanners and 3D printers being properly maintained? Many of the checks will focus on safety but some will report on serviceability. Downtime is waste. How are the robots, laser scanners and 3D printers being checked before operations?

5 When materials are being pumped, is the robot suitably protecting those materials?

6 Has the data for laser scanners, robots and 3D printers been quality assured? Are the scans that are sent to stakeholders fit for purpose to meet agreed specifications? Is the scanned data secured and available as part of the digital asset management?

7 Can robots, laser scanners and 3D printers demonstrate they are within specified calibration requirements? Are there calibration records available prior to the scan being created?

8 Are robots and 3D printer supervisors and laser scanner operators suitably competent? Can training/qualifications be demonstrated?

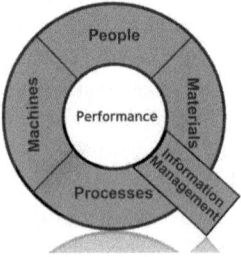

Notes

1 Tybot, 'Reliable, flexible, and scalable solution for bridge deck construction'. Retrieved from www.tybotllc.com

2 HSE, *Health and Safety Statistics for the Construction Sector in Great Britain, 2017* (London: Health and Safety Executive, 2017).

3 FBR, 'Robotic construction is here'. Retrieved from www.fbr.com.au/view/hadrian-x

4 Blueprint robotics, 'A better way to build'. Retrieved from www.blueprint-robotics.com

5 Construction robotics, 'SAM100'. Retrieved from www.construction-robotics.com/sam100/

6 Smisek, P., *A Short History of 'Bricklaying Robots'*. 17 October 2017. B1M video channel. Retrieved from www.theb1m.com/video/a-short-history-of-bricklaying-robots

7 Pathé News, *Mechanical Bricklayer 1967*. 30 April 1967. Retrieved from www.british-pathe.com/video/mechanical-bricklayer

8 NCCR Digital Fabrication, *In situ fabricator*. 18 June 2015. Retrieved from www.youtube.com/watch?v=loFSmJO3Hhk

9 NCCR Digital Fabrication, 'In situ fabricator mesh reinforcement'. 29 June 2017. Retrieved from www.youtube.com/watch?time_continue=29&v=TCJOQkOE69s

10 Brokk Inc., 'The smart power lineup'. Retrieved from www.brokk.com/us/

11 *Now Science News*, 'HRP-5P Humanoid Construction Robot by AIST'. 30 September 2018. Retrieved from www.youtube.com/watch?v=qBvuZ-tUFiA

12 Honda, 'ASIMO'. Retrieved from http://asimo.honda.com

13 Doxel, 'Artificial intelligence for construction productivity'. Retrieved from www.doxel.ai

14 Medium, 'Introducing artificial intelligence for construction productivity'. Retrieved from https://medium.com/@doxel/introducing-artificial-intelligence-for-construction-productivity-38a74bbd6d07 (accessed 24 January 2018).

15 Allied Market Research, 'Construction lasers market by product'. Retrieved from www.alliedmarketresearch.com/construction-lasers-market (accessed September 2018).

16 McDonald, R., *Root Causes and Consequential Cost of Rework* (Catlin Insurance North America Construction, 2015).

17 Leica and Autodesk, 'When to use laser scanning in building construction'. (2015). Retrieved from http://constructrealityxyz.com/test/ebook/LGS_AU_When%20to%20Use%20Laser%20Scanning.pdf

18 3D Print Canal House. Retrieved from http://3dprintcanalhouse.com

19 Dubai Future Foundation, *Office of the Future*. (2018). Retrieved from www.officeofthefuture.ae/#

20 Dubai Future Foundation, 'Dubai 3D printing strategy'. (2018). Retrieved from www.dubaifuture.gov.ae/our-initiatives/dubai-3d-printing-strategy/

21 Van Wijnen, 'World first: living in a 3D printed house made of concrete'. (2018). Retrieved from https://translate.google.com/translate?hl=en&sl=nl&u=www.vanwijnen.nl/actueel/wereldprimeur-wonen-in-een-3d-geprint-huis-van-beton/&prev=search

22 BBC News, 'The world's first family to live in a 3D-printed home'. 6 July. 2018. Retrieved from www.bbc.co.uk/news/technology-44709534

23 Yhnova, 'A robot 3D printer is building a house in Nantes'. (2017). Retrieved from http://batiprint3d.fr/en/

24 *NUS News*, 'NUS builds new 3D printing capabilities, paving the way for construction innovations'. 5 July 2018. Retrieved from https://news.nus.edu.sg/press-releases/construction-3D-printing

25 ICON, 'Welcome to the future of human shelter'. (2018). Retrieved from www.icon-build.com

26 ICON, 'New Story + ICON'. March 2018. Retrieved from www.iconbuild.com/new-story/

27 DFAB House, (2018). Retrieved from http://dfabhouse.ch/dfab-house/

28 MX3D, (2018). Retrieved from https://mx3d.com

29 MX3D, 'MX3D bridge'. September 2018. Retrieved from https://mx3d.com/projects/bridge/

30 Skylar Tibbits, 'The emergence of "4D printing"'. (2013). TED. Retrieved from www.youtube.com/watch?time_continue=1&v=0gMCZFHv9v8

31 Balfour Beatty, 'Building the future with 3D printing'. 16 November 2016. Retrieved from www.youtube.com/watch?time_continue=86&v=EogNa8LAWQg

32 Oak Ridge National Laboratory, 'Project AME'. (2017). Retrieved from https://web.ornl.gov/sci/manufacturing/projectame/

33 ETH Zürich, 'Knitted concrete'. (2018). Retrieved from www.youtube.com/watch?v=spPpkPHK7Q0&feature=youtu.be

34 3Dnatives, 'The 3D printing construction market is booming'. 26 January 2018. Retrieved from www.3dnatives.com/en/3d-printing-construction-240120184/

15 Augmented reality (AR), mixed reality (MR) and virtual reality (VR)

There is a spectrum that goes from the real-world environment to a fully virtual world (VR) with augmented (AR) and mixed realities (MR) sitting in the middle of the spectrum.[1] The differences can be broken down further[2] as the slider moves from the real world towards this extended reality (XR) experience, which is confusing for the average user when trying to decipher between the different technologies.

Augmented reality or AR is a computer-generated view of the world overlaid on the actual world (see Figure 15.1). The user will see, using specialised hardware and software, what is there already and what could be there, from 3D digital content. An AR display allows the user to observe and interact with

Figure 15.1 An augmented reality overlay brings design into the real world.

Source: Patrick Schneider@patrick_schneider.

digital content created by an AR rendering engine and viewed through a content management system. Tracking the location allows the AR environment to sync with the real world. By wearing an AR headset, a user could view a BIM model of a building overlaid on a construction site that has only laid foundations, as they walk around the site, making it much easier to visualise the in situ building and assisting with better quality feedback.

Gartner identified AR as an emerging technology back in 2005 on their Technology Hype Cycle but it struggled to gain traction in saleable products. In particular, the cost of innovative products has been a major stumbling block for uptake by Tier 1 contractors, although several have experimented. In addition, early devices such as Google Glass (which had little application to construction) only allowed digital content to be viewed, with little distinct advantage over looking down at a tablet or smartphone to see the same information.

An AR trial lasting five months undertaken by Crossrail[3] at the Custom House site in 2014, piloted the capture of construction progress information against a BIM model. The information was recorded on site in a 3D AR model and synced with a 4D model to monitor project progress. It demonstrated significant productivity gains from process management, as it reduced the time needed by 73 per cent to update the model against as-built progress. While this small study showed the potential, it also highlighted the constraints of uploading data to the AR hardware (in this case, a tablet) and the accuracy at that time of recording data using the apps available.

In 2018, Microsoft HoloLens, Magic Leap Lightwear, Epson Moverio (shown in Figure 15.2), Google Glass Enterprise Edition (a new business version of the glasses), Vuzix Blade AR, Meta 2, Optinvent Ora-2, Garmin Varia Vision, ODG R-7 Smartglasses have all developed a range of functionality and features. Some, like the Epson Moverio,[4] provide a head-mounted display linked to DJI's drone cameras, making it easier to fully see the images through the 'eyes' of the drone, while maintaining a crucial line of sight. This First Person View (FPV) control is an exciting development by Epson with the smart glasses having six-hour battery life and a built-in front-facing 5MP camera for taking additional photos and videos.

GE Aviation uses UpSkill software with Google Glass Enterprise Edition for assembly line workers, replacing a huge Lever Arch file of work instructions, with the smart glasses.[5] The workers can pull up training videos to view through the glasses and call up remote experts. It takes a little practice and confidence using such glasses for work activities instead of traipsing to an office to find a manual or set of procedures but it becomes intuitive and produces a stream of data on accessibility and usage, to deliver future improvements in the technology. With torque wrenches linked to wi-fi, the worker can be instructed when the jet engine nut has reached the correct position of torque. Then a photo is taken and data recorded of the results using the glasses providing an audit trail in the event of investigations.

DAQRI have focused their business on creating their Smart Glasses[6] product with the WorkSense applications suite, which provide a hands-free, AR experience.

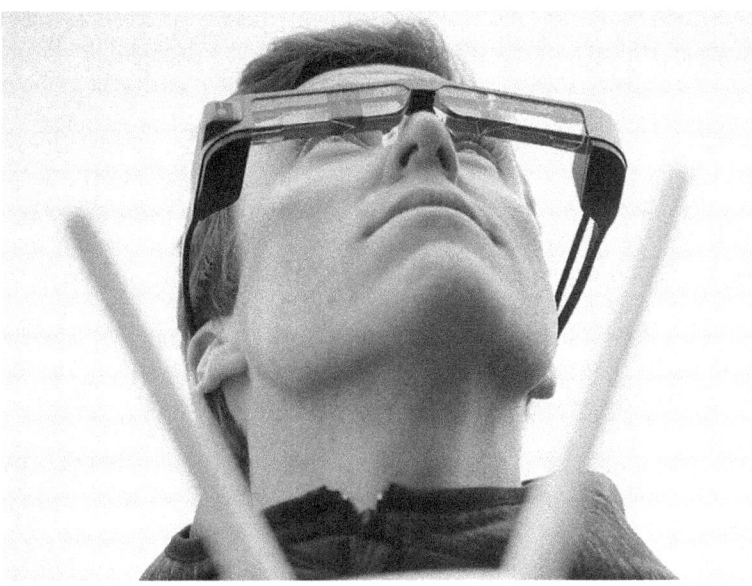

Figure 15.2 Epson Moverio BT-300 smart glasses with First Person View of head up display linked to drone cameras.

Source: Epson.

Eye gaze controls the apps by focusing on buttons that open and close menu options. From extensive customer research, it delivers solutions to five tasks:

1   Showing an issue to someone else to ask for assistance.
2   Scanning the environment to make a record.
3   Tagging elements within that environment or accessing tag data.
4   Modelling designs.
5   A guide app to provide written process steps.

The glasses can be used for trainees to see a new overlaying pipework design on a wall with work instructions to hover in front of the wearer to explain steps in a process, such as opening a valve.

The tag app can be used by a quality professional walking around a construction site and digitally labelling snags for fixing or after construction completion, linking management system documents or performance data to physical assets for operators to easily find. A report could pop up when looking at a tag on a pump that shows the live data for the motor speed, flow rates and suction pressure. The scan app can video the environment in a hands-free way that avoids juggling a camera phone and deters the user from being tempted to take off safety glasses, as they focus their eyes on photos or videos. The author tried out an earlier DAQRI product called the Smart helmet in 2015 that worked in the same way as the glasses but had an integrated safety helmet. It did take a few minutes getting used to gazing at menu options before they would open but it didn't take long to then open up work instructions and follow them on a dummy set-up for controlling a valve on pipework.

AR offers quality professionals an opportunity to design scenarios to overlay the real world, for example, demonstrating to site managers the effects of mistakes in quality management. It can be used as an education and training tool to show team members the impact of failing to plan out inspection and testing routines for construction activities. To do that, the quality professional needs to appreciate what AR can achieve and to commission or lobby for funding to create demonstrations. Sitting back and hoping such AR technology may be created is folly. We need to be proactive, vocal and leaders in advocating for such tools that will burnish our reputation as being in the vanguard of technological advances.

Magic Leap is a product that has received a lot of investment and media attention, and only launched in 2018.[7] It produces a head-mounted, virtual retinal display that projects a light display into the user's eye. Using a version of AR that it calls 'Mixed Reality' (MR) means Magic Leap can produce realistic digital images for gaming of, say, waterfalls and celestial planets blended into the natural light floating in front of you through six cameras mounted on the goggles. However, the high resolution and believability could be a step in construction towards viewing much more life-like supplier parts that can be turned upside down and inside out before ordering, rather than being viewed as flat 2D images on a website.

Virtual Reality or VR started as a gaming product to immerse players in their games rather than just see the action on their desktop screens and televisions. The difference with AR is that VR takes you into a full artificial world rather than just overlaying it, on what we see around us. However, it soon began to have uses other than gaming and has crept across into business. Headsets are either mobile or tethered, with the mobile versions allowing either a smartphone to be inserted into them or standalone VR headsets that have in effect their own Android phone hardware built in.

The leading business VR headsets in 2018 are HTC Vive and the Oculus Go (and Rift) with Samsung Gear VR and Google Daydream View close behind. The technology, though, is developing rapidly with the launch of Oculus Quest providing greater immersive experiences and freedom with the absence of cabling. Hand-held motion sensors allow the user to point to a spot in the digital world and click to open up options to change the representation.

For construction, VR provides a full digital representation of a design that the user can look and walk around, true to scale. In the past, a physical model would have been built out of balsa wood to show a client a mock-up of an interior. The problem is that these are slow to create and inefficient to change, when the client provides feedback on what they want. VR can be updated and changed instantly as the user can point and shoot motion sensors to change colours, textures, materials and the layout of a virtual representation. ariot.io,[8] based in Austria, can automatically convert a BIM model into an AR environment and can pull in Internet of Things (IoT) data from sensors to enrich the visualisation.

There are some cautionary words of advice; the experience can be confusing and disorientating to the point of feeling nauseous and dizzy. If a user has a disposition towards seizures, then they should seek medical advice before trying on a VR headset. As with many IT products, eye strain and headaches can be brought on from prolonged usage.

For H&S training, VR can place someone in a potentially harmful environment but with very low risk. Without leaving the classroom, they can be taught to identify tripping hazards or how to work safely at height. Workers can be trained in emergencies to better cope with fires, scaffolding collapses or vehicle collisions on site. In business continuity, teaching staff what to do in a variety of low probability but high severity situations can make the difference between life and death but few businesses will invest in such rare environments (until too late).

VR can be a tool in enhancing customer satisfaction in quality management. End users can experience and provide feedback on designed built environments. Nurses can enter the VR world and walk around a new hospital ward and explain to designers preferred changes to layout where immersion can bring out subtle problems that may not be noticed on drawings or even in fly-throughs of 3D digital models.

When Layton Construction used this technology to design a hospital in Florence, Alabama, they found that mock-up construction costs were reduced by 90 per cent,[9] and greatly increased satisfaction from end-users. As an example, nurses identified that oxygen canisters designed to be located in delivery rooms would have blocked wall outlets. The feedback allowed both design changes and prefabrication of new components saving money. 3D Dynamic VR by 3D Repo uses VR glasses to provide real-time experience of BIM models with visualisation of construction sequences to enhance understanding by construction professionals of activities and their risks.[10] It can also be used to create 3D VR worlds for end-users like the public, to understand and be consulted on construction projects.

Quality professionals could use VR for training to raise awareness of poor material delivery, movement and storage routines or training to carry out humble concrete slump and cube tests. For better interactive experiences, bringing people together through shared VR will enhance problem solving abilities but it is currently a difficult business case to justify the cost versus the benefit.

HoloBuilder is a start-up that captures 'walk-throughs' of construction site with geotagging of snagging (punch lists) items of defects and re-work required.[11] Vision technology allow distance measurements to be made and off-site collaboration, with stakeholders who may be located miles away. Remote quality professionals can be consulted from their desks, without necessarily visiting the site to discuss quality control problems.

In the absence of a BIM model, the Magicplan app is a spectacular AR technology that takes photos of an existing building and instantly converts them into basic floor plans with detailed measurements. [12] Using a smartphone or tablet, the app allows geotagging of notes and images to highlight defects, create operating instructions and list queries. Materials and activities can be calculated and ordered with a few clicks. For smaller building and maintenance companies, it is the sort of technology that is simple to use, competitively priced and aids productivity and quality improvements in business processes.

Where VR has been useful in Lean Six Sigma (LSS) and TRIZ training,[13] it has been used with the cheaper type of VR headsets, such as Google Cardboard that allows any smartphone to be slotted into them with some basic virtual reality experience.

ResearchAndMarkets.com have estimated that the AR and VR market will grow globally to $94.4 billion by 2023,[14] which will bring with it cheaper prices and better functionality, making it highly likely that such technology will penetrate into construction quality management.

Quality professionals need to be much more vocal in demanding access to the latest AR/MR/VR technologies and set out robust business cases of why they can reduce risk and improve quality. The more operatives on site (not just quality professionals) who can compare and contrast what is being built, with what has been designed, the better the chances quality outcomes will improve. For that, we need to raise our game and appreciate that the old ways will never persuade senior decision-makers to invest in Digital Quality Management.

To prove such case scenarios where reality technologies become commonplace, we need data to demonstrate the value. That means recording not just case studies of finding a misalignment in, say, an AR viewing, between the design and the as-built, but also ensuring the numbers and types of such issues are recorded. That allows the Cost of Quality calculations to be made as Rosendin Electric found in one incident that avoided expensive re-work.[15] We must become much more assiduous in raising expectations that quality issues found using technology are formally recorded to allow measurement analysis and root causes to be reported. Otherwise, the data is lost and the evidence becomes subjective.

To get the most from AR technology, a 5G network will make a significant difference, with speeds of 100 times what we can currently get on the mobile networks. The fifth generation mobile network system is the next step change improvement after 2G (texts and photo messaging), 3G (video calling and mobile data), and 4G (gaming and video streaming) but as usual will be typically first put out to major cities and urban areas, leaving rural parts of countries left waiting. It is expected that the USA, Japan, South Korea and China will be leading the roll-out in 2019 with the UK trailing in 2020. 5G will mean that walking around site using smart glasses connected by wi-fi will become almost instantaneous in data transfer, uploads and downloads, making the business use much less painful and more attractive.

Digital learning points

Box 15.1 Digital learning points: augmented reality (AR), mixed reality (MR) and virtual reality (VR)

1   Augmented reality (AR) has a great capacity to add functionality and insight into the as-built world vs the designed BIM model, identifying errors and defects and providing Cost of Quality to be calculated.

2   Combining technologies such as AR and smart glasses allow eye gaze to call up work instructions on site and in the factory. Check that such QMS 'documents' maintain version control.

3   Virtual reality (VR) offers simulated Quality Management training such as identifying snagging. So, identify scenarios for added value training and find cost-effective suppliers. Lobby to develop such training.

4   Promote VR for clients to view designs and construction scenarios and measure customer satisfaction, before and after.

5   AR, MR and VR smart glasses will become much more efficient and attractive to use in business, with the introduction of the new 5G mobile network.

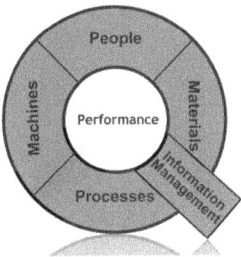

Notes

1   Milgram, P., Takemura, H., Utsumi, A. and Kishino, F., 'Augmented Reality: A class of displays on the reality-virtuality continuum'. (1994). Retrieved from http://etclab.mie.utoronto.ca/publication/1994/Milgram_Takemura_SPIE1994.pdf

2   Liu, J., 'The difference between AR, VR, MR, XR and how to tell them apart'. Hackernoon, 2 April 2018. Retrieved from https://hackernoon.com/the-difference-between-ar-vr-mr-xr-and-how-to-tell-them-apart-45d76e7fd50

3   Crossrail Learning Legacy, 'Augmented Reality trials at Crossrail'. 14 March. 2017. Retrieved from https://learninglegacy.crossrail.co.uk/documents/augmented-reality-trials-crossrail/

4   Epson, 'A bright horizon for FPV'. Retrieved from www.epson.co.uk/products/see-through-mobile-viewer/moverio-bt-300/drone-piloting-accessory

5   Kloberdanz, K., 'Smart specs: OK glass, fix this jet engine'. *GE Aviation*. 19 July 2017. Retrieved from www.ge.com/reports/smart-specs-ok-glass-fix-jet-engine/

6   DAQRI, 'Smart glasses'. Retrieved from https://daqri.com/products/smart-glasses/

7   Magic Leap, 'Free your mind'. Retrieved from www.magicleap.com

8   ariot.io, 'Finally, Augmented Reality for the building lifecycle'. Retrieved from www.ariot.io

9 Angus, W. and Stocking, L.S., 'VR transforms doctors, nurses, and staff into virtual construction allies'. Autodesk. 1 November 2017. Retrieved from www.autodesk. com/redshift/vr-construction/

10 3D Repo, 'Dynamic virtual reality for customer engagement and staff training in construction'. November 2017. Retrieved from http://3drepo.org/wp-content/ uploads/2017/11/Dynamic-VR_a4booklet.pdf

11 HoloBuilder, 'Full HoloBuilder feature overview'. Retrieved from www.holobuilder. com/features/

12 Magic-plan app, 'Create a floor plan within seconds'. Retrieved from www.magic-plan.com

13 ASQ Kaushik, S.K.V., 'Virtual reality for quality'. June 2017. Retrieved from http:// asq.org/2017/06/lean/virtual-reality-vr-for-quality.pdf

14 CISION PR Newswire, 'Global Augmented Reality (AR) and Virtual Reality (VR) market is forecast to reach $94.4 billion by 2023 – soaring demand for AR & VR in the retail & e-commerce sectors'. 31 July 2018. Retrieved from www.prnewswire.com/ news-releases/global-augmented-reality-ar-and-virtual-reality-vr-market-is-forecast-to-reach-94-4-billion-by-2023-soaring-demand-for-ar-vr-in-the-retail-e-commerce-sectors-300689154.html

15 EC&M, '3D visualization brings a new view to job sites'. 1 October 2018. Retrieved from www.ecmweb.com/neca-show-coverage/3d-visualization-brings-new-view-job-sites

16 Wearable and voice-controlled technology

The quality professional needs to embrace, thrive on and master technology, as a precursor to improving quality and efficiency of their profession. Until very recently the quality professional could only typically access digital information using fingers typing on a keyboard, clicking a mouse or touching a screen. It is a relatively slow, inefficient process and glues the operator to a laptop, tablet or desktop computer. By being able to use other senses and motions, the options and potential for productivity are increased. Our voices and gestures enrich our ability to perform at work. Interacting with narrow artificial intelligent machines is being actively used in offices and construction sites and allows quality professionals to increase productivity.

Think about Amazon's Alexa or Google Home and asking it to find and play a piece of music. Done within a second. How much quicker and easier is that compared to getting up walking to a shelf and searching for a CD, opening the case, opening the CD player door, entering the CD, closing the door, pressing the play button and altering the volume before returning to the seat? Yet that is what we frequently do in an office to find a Lever Arch file, book, report or other piece of hard formatted information. Using word search in a folder or document can be better (but depends on the folder ontology and whether the information actually resides within the server or hard drive being used) than the manual retrieval of paper but how much time is wasted trying to find the right information at the right time? Finding the right information or high quality and useful information, is key to reducing errors and increasing good decision-making in quality management.

Voice control is becoming widely used in the home for a wide variety of information tasks, such as establishing the weather, playing music and creating to-do lists. In the professional sphere, voice is used much less (probably due to the regimented routines and peer pressure conformance rather than anything to do with open plan office environments, since people speak on the phone and to each other all day). Yet asking for the retrieval of a document, while other duties are being performed saves time and maintains concentration.

Gartner previously predicted 30 per cent of our interactions with technology would be through voice control by 2018, but it is likely to take longer in business. Nevertheless, quality professionals need to explore the opportunities within the range of voice-controlled AI products.[1]

There are a staggering 20,000 smart home devices[2] compatible with Alexa used by 3,500 brands with 10,000 new Alexa skills being added every four months.[3] As world leaders in adopting voice assistants, over a quarter of the US population[4] uses voice technology in connecting with their smartphones.

As the home device IoT market of voice-controlled technology becomes extensive, so the migration to business equipment and devices is likely to take off. Whether commanding robots, laser scanning devices or 3D printers, we should be prepared to cast off our inhibitions of speaking to machines (and accept they will get it wrong sometimes). IBM's Watson's Assistant has been proclaimed to be an AI market leader for business with a quickly adaptive approach to understanding the changes in business terminology and complex situations.[5]

'Alexa for Business' can provide the conduit for queries on simple tasks such as meeting room bookings, making phone calls, translation, dictation of reports and managing calendars.[6] Conference calls can be made through Alexa to multiple participants using their names instead of trying to dial different telephone numbers. How many times have you seen a queue of people waiting impatiently outside a meeting room, as someone else is inside? Then one or two go marching off in different directions trying to find out where they should be to speak to IT or administrators or the poor bloke sitting next to the meeting room. The mini-chaos wastes so much time. Instead, asking the technology may get answers more readily. It can also be trained to understand business jargon and the voice recognition notes who is speaking and what they are saying. Some people will get twitchy at privacy issues but this is the world we live in and we should exploit the technology for every penny its worth to make ourselves and our businesses better.

When voice control becomes difficult, either due to the need for a quiet environment or, conversely, in a too noisy environment, then gestural control comes into its own. The MYO wristband allows head up pages to be turned with a flick of the finger and a drone's elevation to rise by a gentle move of the arm.

Much of quality control relies on the quality professional assessing the partially or fully completed product or component against an agreed specification. It may be a brick wall being built or bolted steel beams that require visual inspection. Easy to do, some may suggest, when the weather is fine and there is little pressure but that is not typical of construction. The weather may be cold, the pressure may be high and physical and mental challenges may result in mistakes.

Eye-tracking eye wear such as Tobii Pro[7] allows cameras on glasses to track the wearer's gaze. The data capture of the time of their attention and where they are looking is fed back to researchers who can see and hear what the user saw and heard in a digital format to provide real insights without relying on someone feeding back by talking and describing what they saw (with all its inherent flaws and bias). It can be used in root cause investigations to assess situational awareness reasons for missing quality defects in components before being fitted into construction or confirming the user misread measuring equipment.

The data can reveal work environment improvements, such as wasted time in moving materials on site and spot human factors influencing performance by revealing gaze-plots and heat maps comparing where a trainee is looking versus a seasoned professional. Such technology is very useful for assessing and teaching quality professionals so that knowledge can be passed on from those who have more experience to those who are learning. Such training can then be carried out in a VR environment or on site.

A number of the top Tier 1 UK contractors have used and proved the augmented reality technology from DAQRI products (as described in Chapter 14), such as their hard hats that meet British Standards for safety and have a visor for head up displays of information and data. Although the hard hats have been discontinued in production, the standalone smart glasses are available with WorkSense software. The glasses overlay a work environment using AR with the ability to tag defects seen on site and match BIM models to actual build to compare and contrast.

Similarly Boeing uses SkyLight smart glasses by Upskill in their production lines.[8] In the past, the workers would have had to compare drawings on a laptop with the bundles of wires in front of them to identify which wires enter which connectors. With sometimes up to 100 wires entering a single connector, the concentration needed to switch between looking at a screen and operating a keyboard, and moving back to the production line, invariably slows down the process and increases the likelihood of mistakes. In a Boeing 737 airplane, mistakes in production are unacceptable; they can cost lives. By using the smart glasses, the worker can keep both hands free by using voice control while they work and since the glasses can also stream the video images seen by the worker, supervisors can be called up for advice and they can log into their laptops to see the same images. Boeing has measured the productivity improvements at cutting production time by 25 per cent and lowering error rates to nearly zero. With design for manufacture and assembly (DfMA) processes, there will be greater use of off-site pre-fabrication, increasing the analogies with manufacturing technology. Checking out now how such technology can improve construction quality management will give the profession a head start.

Epson Moverio smart glasses are used in conjunction with drones to enhance the capability of the operator.[9] The glasses can be worn over prescription glasses or prescription lenses can even be fitted to the smart glasses. The AR image provides the operator with a full head up display that provides data that can usually be seen on a smartphone or tablet using an app but the glasses allow the operator to keep their head up, looking at both the drone in the sky and the HUD showing the camera images and the data. Overlaying range, height, record button, etc. on the HUD with a mouse controlled by a hand-held unit means the operator avoids constantly bobbing their head up and down and provides a better user experience and reduces risk of losing line of sight of the drone.

XOne smart glasses by XOEye Technologies have a range of features geared towards industry use:[10]

- microphones and speakers for two-way audio communication and collaboration;
- barcode scanning;
- a suite of sensors that include biometrics, accelerometer and gyroscope for data measurement.

In 1982, the Seiko TV Watch, that could receive 82 VHF and UHF channels and FM radio, was launched. It could transmit television pictures to a tiny screen but had to be plugged into a box of electronics that had to be carried around with it. In the late 1990s, basic smart watches began to appear in the marketplace with the Seiko wrist computer called MessageWatch and a first phone watch, Wristomo, by Nippon Telegraph and Telephone Corp. in Japan. These did not capture the imagination outside of geeks who had waited decades for a 'Dick Tracy' watch. In 2003, Microsoft brought forward its Smart Personal Object Technology (SPOT) for watches but again it rapidly faded away. It was perhaps the Pebble watch in 2013 that seriously excited consumers but over the next couple of years it suffered adverse publicity and although purchased by Fitbit, also disappeared from the market. Since then, Fitbit has carved out a market segment with their fitness trackers and smart watch products.

It was Apple that developed a sustainable smart watch product in 2014, which has seen year-on-year increases in sales to around twenty million per year in 2018 and it is expected to increase to thirty-one million by 2020; nearly 50 per cent of the global smart watch market. Aside from Apple watches taking and making calls, playing games, monitoring heart rates, hard fall detection and SOS calling and hundreds of apps (as well as telling the time), the smart watch has other great potential for industrial settings.

The WorkerBase smart watch has data acquisition capability through a QR code scanner and an 8 megapixel camera, with dozens of manufacturing apps.[11] It can send text and audio messages to workers that there are missing parts in a component build unit and provide simple work instructions, such as carrying out scheduled tests on equipment using workflow management software. Upon confirmation that one operator has finished a task, a message can automatically be sent to a quality professional to carry out an inspection with a full digital audit trail and capability to take photos. The detailed reports showing time, cost and quality can be a rich source of data analysis to develop improvements in processes.

Smart construction clothing has slowly emerged with prototypes aimed particularly at health and safety enhancements. SolePower has designed self-charging, smart safety boots that can measure user fatigue and falls, can light up to improve visibility, have location sensors to send safety alerts, if the user strays into geofenced plant zones and display temperature for warning of heat exhaustion or excess cold weather.[12]

GuardHat is a hard safety hat for operatives that has an RFID reader for proximity readings, an accelerometer to track sudden falls and a carbon monoxide gas monitor.[13] It has three buttons: audio, video and an SOS alert for communications, and meets all required international standards of manufacture.

Even basic clip-on sensors attached to belts provide useful data that can make the difference to workers suffering falls on site. The Spot-R clip automates attendance times on site, automatically warns supervisors of the location of a hard fall and the inbuilt alarm can aid site evacuation in case of emergency.[14] Data is only collected while on the designated site.

For quality management, the premise of knowing the location of people and machines may be useful for root cause analysis, after a quality control incident or non-conformance but specific apps need to be designed and available commercially that will provide the data visualisation in a format that we can use easily. Again, as a profession, we need to be creatively developing ideas for practical application by manufacturers, to generate greater interest in the quality management marketplace.

The German company, ProGlove, has a range of smart gloves for different industrial applications.[15] A small electronic box is fitted to the back of the gloves, allowing bar code scanning without the handheld scanning 'pistol' often seen inside factories. It can provide feedback to operators through flashing lights, haptic and acoustic signals. Lufthansa use ProGlove to assist with material handling of 100,000 parts in their warehouses. The complex supply chain can be tracked at every step in the process from leaving the supplier to arrival in a warehouse and storage shelf, to despatch and to its final destination using the scanner on the back of the glove. It can even alert the operator to new priority deliveries needing to be couriered. It saves looking around for a scanner and improves productivity. Such devices could be routinely deployed on a construction site to improve handling of materials.

With all these products the users' data must be meticulously protected and preserved under data protection regulations and to ensure appropriate levels of privacy. Ideally, all workers would be issued with basic digital devices inbuilt into personal protective equipment (PPE) but there may need to be careful consultation and education on how, when and why the data is being collected by companies. If insurance companies heap pressure on contractors to make workers wear these devices, by offering lower premiums and employment contracts continue to require more personal data to be held, then workers may find that opting out is difficult. The onus then is on businesses to be far more robust in protecting their workers' data to maintain strict confidentiality and information security. Before such devices are commonplace at construction sites and off-site manufacture, the price point will need to fall substantially and the benefits need to be tested and proved.

Exoskeletons by EksoBionics offer a solution to both construction worker injuries and greater productivity.[16] The EksoVest is an upper body exoskeleton that helps to elevate and supports an operator's arms, to assist them with chest height and overhead tasks. Tools such as grinders and rotary hammers become weightless while wearing the supportive exoskeleton. With less fatigue, the risk of injury is lower and the level of workmanship is likely to improve. Not every construction worker needs this device but for some tasks, it is an obvious enhancement. However, a next step in development will be to improve the data feedback to allow improvements in its design and the application of effective construction processes.

Likewise, Honda, as a spin-off from the ASIMO robot technology, have created a range of personal mobility assistance devices, from walking assist to bodyweight support, which have industrial uses.[17] After ten years of muted publicity, Honda's walking assist device that is used in 250 facilities in Japan has secured EU medical device approval for CE marking,[18] which provides it with an export passport to Europe.

While many wearable technology products are geared towards safety, the quality management profession needs to collaborate and identify functions and uses that will add more value to our discipline in order to push manufacturers to design such products and construction companies to adopt their usage. Health and safety concerns give these products a way into everyday use in the industry but we need to be proactive in identifying how quality control inspection and testing can be improved, as well as quality management through wearable clothing that aids training and accessing management systems.

Digital learning points

Box 16.1 Digital learning points: wearable and voice-controlled technology

1   Experiment with wearable and voice-controlled technology to improve efficiency and effectiveness in the workplace. Consider using IBM Watson Assistant or Alexa for Business.

2   Use eye-tracking eye wear, such as Tobii Pro, to research and develop training in quality management. Which defects or errors have been missed?

3   Investigate uses for smart glasses by DAQRI to test out accessing management system 'documents' and tagging defects and queries.

4   Think about comfort and ease of use of this technology, especially over a long day. Ensure processes for optimising the use are thought out and documented. Eye strain from smart glasses may become the new 'repetitive injury' in coming years.

5   What can the quality profession learn from manufacturing on the application of wearables? Smart glasses, watches, gloves and exoskeletons have industrial uses and we need to assess and develop them to exploit quality management capability.

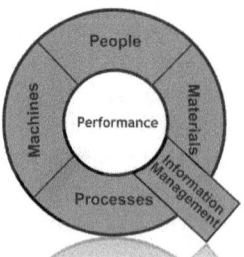

Notes

1 Gartner, 'Market trends: voice as a UI on consumer devices — what do users want?' 2 April 2015. Retrieved from www.gartner.com/doc/3021226/market-trends-voice-ui-consumer

2 Karczewski. T., 'IFA 2018: Alexa devices continue expansion into new categories and use cases'. Alexa Blogs. 2 September 2018. Retrieved from https://developer.amazon.com/blogs/alexa/post/85354e2f-2007-41c6-b946-5a73784bc5f3/ifa-2018-alexa-devices-continue-expansion-into-new-categories-and-use-cases

3 Kinsella, B., 'Amazon Alexa now has 50,000 skills worldwide, works with 20,000 devices, used by 3,500 brands'. Voicebot.ai. 2 September 2018. Retrieved from https://voicebot.ai/2018/09/02/amazon-alexa-now-has-50000-skills-worldwide-is-on-20000-devices-used-by-3500-brands/

4 eMarketeer, 'Alexa, say what?! Voice-enabled speaker usage to grow nearly 130% this year'. 8 May 2017. Retrieved from www.emarketer.com/Article/Alexa-Say-What-Voice-Enabled-Speaker-Usage-Grow-Nearly-130-This-Year/1015812

5 IBM, 'Watson Assistant'. Retrieved from www.ibm.com/watson/ai-assistant/ (accessed 12 April 2018).

6 AWS, 'Alexa for Business'. Retrieved from https://aws.amazon.com/alexaforbusiness/

7 Tobii Pro, 'Eye tracking for research'. Retrieved from www.tobiipro.com

8 Upskill, 'Augmented reality use cases in enterprise'. Retrieved from https://upskill.io/skylight/use-cases/

9 Epson Moverio, 'DJI MAVIC AIR with Epson Moverio BT300 smart glasses – review part 2 – flight test, pros & cons'. 28 March 2018. Retrieved from www.youtube.com/watch?v=7u5NvNOdSeg

10 XOi Technologies, 'It starts in the field'. Retrieved from www.xoi.io/solution/

11 WorkerBase, Goods inspection app. Retrieved from https://workerbase.com/industry-software/goods-inspection-app/

12 SolePower, Smartboots. Retrieved from www.solepowertech.com/#smartbootsgraphic

13 GuardHat, 'Overprotective in a good sort of way'. Retrieved from www.guardhat.com

14 Triax, spot-R. Retrieved from www.triaxtec.com/workersafety/how-it-works/

15 ProGlove, 'Ready for 4.0'. Retrieved from www.proglove.de/products/#wearables

16 EksoBionics, 'Power without pain'. Retrieved from https://eksobionics.com/eksoworks/

17 Honda, 'ASIMO innovations'. Retrieved from http://asimo.honda.com/innovations/

18 Honda, 'Honda walking assist obtains medical device approval in the EU'. 18 January 2018. Retrieved from https://world.honda.com/news/2018/p180118eng.html

17 Blockchains

Traditionally, transactions such as payments or contractual agreements are maintained through a ledger that each party keeps. Blockchains are a digital ledger of transactions (see Figure 17.1). The problem is that through each separate copy, the potential for fraud or human error increases, whereas a blockchain is a single source of truth, with full audit trail. Time and date stamps record each transaction, which is identified by their unique cryptographic number, so changes can be fully verified, are irrevocable and groups of related transactions are then kept in blocks. Costs are low, as there is no manual, human activities within the process for verification and reconciliation.

This single version is distributed across millions of servers and is routinely monitored to check on accuracy. Since it is not held centrally, it makes it impossible (so far) to hack and change, since the routine checks every ten

Figure 17.1 Blockchain graphic to represent the digital ledger of transactions.
Source: Pete Linforth, https://pixabay.com/en/users/TheDigitalArtist-202249/.

minutes or so, it would pick up on any such amendments on one server that are not reflected across all the servers. Transactions become almost instantaneous as the ledgers update simultaneously.

Currently, banks maintain security by locking down versions of a transaction at each end and only allowing a change by unfreezing one end before it can be amended and then copying it to another bank, who similarly wait for it to be unfrozen before they can update their ledger.

Blockchains cut out the intermediary, such as a lawyer or bank, and each party to the transaction can access it directly and update it, as required but each version is strictly maintained. The technology started in 2008 with Bitcoins and other cryptocurrencies and has proved to be unbreakable. Any losses that are reported in the media come from breaches in digital wallets storing the Bitcoins and not the blockchain technology. It is best to separate out all the media talk of Bitcoins and see the underlying technology of trusted transactions, i.e. block-chains as the innovation that the construction industry can use.

Blockchains can empower the use of smarter contracts and payments. For construction, blockchains in the supply chain management could become a powerful method of dealing with the perennial issue of late payments. In the UK, various political administrations over time have attempted to clamp down on the delays experienced by many suppliers and subcontractors, after they have delivered services to a main contractor or client.

In 1996, the Housing Grants, Construction and Regeneration Act was introduced, followed by the Late Payment of Commercial Debts (Interest) Act 1998. The Prompt Payment Code was created by the UK government in 2008 and the Late Payment of Commercial Debts Regulations came into effect in 2013.

In 2014, the government introduced the Construction Supply Chain Payment Charter[1] (which was updated in 2016 and 2018) but only forty-nine companies had actually signed up to the charter, including just three Tier 1 contractors at the time of writing this book. The impact of such a voluntary charter on top of all the legislation is highly questionable, given the objective of faster payments that should be made within 30 days and yet the reality is 42 days on average, up from 40 days five years ago.[2]

The government's long-term plan for the construction industry, 'Construction 2025', made supply chain payments a key goal with the need to 'create conditions for construction supply chains to thrive by addressing access to finance and payment practices'.[3]

Clearly both the mandated and voluntary approaches are failing. This is where the blockchain may form part of the solution. Take a plumber who works for a main contractor on an apartment building. A contract would be drawn up for payments with a retention of, say, 5 per cent kept back to give an incentive for the subcontractor to carry out any snagging and ensure that the main contractor is fully satisfied with the work. However, often in spite of the work being inspected and approved, the invoices languish in the commercial and finance departments of the main contractor. The subcontractor (in this case, the plumber) will send endless emails and make telephone calls, asking for an update on payment.

A huge amount of time is wasted with all these suppliers and subcontractors chasing payments, due to the main contractor holding back to improve their cashflow. Of course, cashflow for the supply chain suffers and can force companies out of business, as they still go on incurring costs on other jobs and paying staff but their invoices remain unpaid. In the age of austerity, banks can be unsympathetic to extending credit lines.

Even if retentions were to be scrapped, and some organisations including Build UK, the Civil Engineering Contractors Association and the Construction Products Association have proposed ending them in the UK by 2025,[4] it still does not tackle the ongoing problem of work that has been satisfactorily completed but not paid on time. Spurious excuses will include new or trivial defects being found, time delays while invoices are being chased up internally or the authorised person being away on sick/holiday/left the business/away from the phone, delaying the release of the payment.

Instead of the long, inefficient and unfair process experienced every day across the UK construction industry, a blockchain would create a contract between the plumber and main contractor and as soon as the inspection was approved, it would release payment. No more paper invoicing and endless delays. The inspection can be carried out by a site manager or site engineer but the implication is that the quality professional may also become an integral part of the process. They will be the ones in the frontline of inspecting work or overseeing the inspection. As such, understanding blockchains will be important to the quality professional. It means that inspections become even more important and there may be pressure from above to reject, based on financial considerations rather than quality management ones. This is where the quality professional's role, credibility and status in a business structure will be tested.

We need construction businesses to give us such authority whereby those who are one step removed from the actual process of construction can independently check on work against specification and pass judgement that senior managers will support and trust. But it also can improve the overall quality management of a business. Instead of allowing poor quality work to take place, businesses have a vested interest in developing quality assurance processes overseeing the supply chain, that will robustly develop partnerships and increase the reliability that the supply chain will deliver the quality required, since the impact will not just be on longer-term costs but also on short-term cashflow as well.

However, does the quality professional need to understand the blockchain technology very well in order to provide an audited, or inspected, level of trust? At the very least, being able to take evidence of the certified existence and acceptance of the blockchain, may be required by the quality professional to create a different, independent audit trail from the site line management.

Blockchains will be a positive development in the industry but businesses need to review their processes to identify how they need to adapt to the different conditions. There is also a valid question of who goes first? In isolation, the blockchain will have limited use until clients and Tier 1 contractors begin to drive its adoption, and it is hard to see it being widely taken up. However, help may be at hand in related and associated sectors.

Take real estate as an example. In Sweden, the land registry authority, Lantmäteriet, with support from the Kairos Future consultancy, and in collaboration with Chromaway AB, Telia Co. AB telecoms and two Swedish banks, SBAB! and Landshypotek, has introduced blockchain technology that it claims could save up to €100 million ($106 million) a year by ending paperwork, tackling fraud, and making transactions much faster. Currently, the signing of a purchase contract through to registration of the sale can be between three and six months. Both the buyer and seller will typically attend an agent's office, to physically sign the documents.

When a land title changes hands, each step of the slow manual process is replaced with a record in the blockchain. The buyer and seller may not even need to be in the same country for the digital signatures to be verified and accepted within the blockchain. It is hoped that with greater acceptance of digital signatures and increasing partnerships with the Swedish tax authority, the ability to use blockchain technology by Lantmäteriet will grow and spread. Land registry is big business with the title-insurance companies in the USA worth more than $2,000 per property, making the sector valued at $15 billion per year.[5]

In the 1970s, Electronic Data Interchange (EDI) was being heralded as the way to speed up payments and protect against fraud since computer-to-computer transactions would be a way to scrap paper invoices. Likewise the ISO 15926 data interoperability standard series has been suggested as a means to secure the data handover verification process in asset management. With these and other approaches, the problem has been the poor take-up rates specifically within the construction industry, which seems to typically recoil against change. Likewise, smart contracts will have a big impact on Joint Contracts Tribunal (JCT) and New Engineering Contract (NEC) approaches, requiring a major overhaul.

Without the widespread adoption of new processes, contracts and standards, then the long supply chain tail pushes back against anything new. Blockchains can certainly work and are an improvement on the current system failures but it will take a robust and determined client and main contractor to implement and drive them forward. We have seen with BIM that clients are invariably unsure of what they want in terms of asset management information and the bulk of the heavy lifting of BIM implementation falls upon main contractors who see cost, time and quality improvements. It is likely that again the Tier 1 contractors will need to be at the forefront of blockchain adoption to make the technology succeed.

Payments can be linked to the BIM model so that as each piece of specified work within a smart contract is completed and approved, so it releases payment, rather than wait for larger pieces of work to be completed or invoicing based on arbitrary timescales, e.g. monthly. If payment is made in cryptocurrencies, then the client, main contractor and each member of the supply chain would need their own digital wallet.

Blockchains could also be an invaluable technology in reliably identifying the source in verified audits of the supply chain of materials. Whether tracing timber back to an individual tree in China or bricks from a supplier in Denmark, the blockchain can demonstrate lifecycle traceability. Trying to certify the origins of materials and maintain tracking, through a myriad of suppliers around the globe, is daunting and open to information delays or breaks that undermine verification. With sealed IoT sensors inside secure tags, then the materials, partial/finished goods can be tracked to final delivery with each transaction locked into the blockchain. Sensitive goods or components can even have sensors tracking temperature and vibration to ensure compliance with rigorous transportation standards. Maintaining such sensors in the tags will be a challenge and quality professionals will need to be satisfied with the way they have been calibrated and protected during transportation and delivery.

The practicality of using such technology and the importance of specific goods and components are likely to take precedence at the beginning, given the expense and level of risk. A heat map of pain points through a supply chain would highlight where the IoT and blockchain technology will add the most value, and can identify situations when it would provide a higher level of protection that a certification is accurate and valid.

Brickschain Foundry[6] has developed a platform for using blockchain technology in the construction industry, without disrupting existing systems that a business already uses, through a single api application. Industry Foundation Classes (IFC)[7] of data can be pulled into a blockchain securely, allowing reports to be run, which can be trusted, since it cannot be changed. It also ensures data interoperability and a chronological reporting of data, which cannot be altered in time through the building management process. For example, if a query is generated on installing an air-conditioning unit, then that query is embedded into the blockchain. Even if it is deleted in another program, then the blockchain retains a copy stating who raised it and when. In the same way, a file can be uploaded to a cloud system and, providing it is being monitored by Brickschain, information will be logged on that upload, including even a hyperlink to find the file.

A quality control snag sheet or punch list, as it is being updated inside a smart sheet, will have that specific item update captured within the blockchain forever. The power that gives a quality professional is immense. It means that there is proof that the snagging list has been updated by a specified user at a specified time and date. It may be by a supervisor on site going through finishing of a room in a house or office. That snagging form cannot be lost or altered at a later date without an original copy always being stored. It increases accountability and transparency in construction work.

For any legal disputes or audits, this is a true record in one place that gives confidence over other systems. It means that instead of submitting evidence to a client on that query being generated and the main contractor then having to double check its authenticity, both parties can trust that indeed the query was raised at the time and on the day claimed. It removes the need and time for these checks and creates confidence.

BuildCoin Foundation is based in Zug, Switzerland, and is a 'non-profit blockchain ecosystem' which the city of São Paulo in Brazil is planning to use, to pay engineers across the globe for feasibility studies in large construction projects.[8] An engineer or subject matter expert (SME) in France may receive an invitation to take part and, upon acceptance, be paid in BuildCoin for her services. The community then votes her work up or down that provides crowd feedback, which assists others in deciding which engineers to use on other projects. For this to be a success, the cryptocurrency needs to be widely accepted so that the engineer can use the digital money. It means that São Paulo can pay for the services without resorting to taxpayer money but it creates a significant risk for subject matter experts on whether payment in this form will actually result in a means that can be used again to make purchases. On the other hand, since it is payment for a feasibility study, then if the project gets a green light, the status and profile of the SME rise and their opportunities for further work also increase.

Blockchains L.L.C.,[9] run by Jeff and David Berns, has bought 67,000 acres of land for $170 million at the Tahoe-Reno Industrial Center in the USA, with the aim of constructing a new 'smart city', east of Reno.[10] The plans include building apartments, houses, shops, schools and offices using blockchain technology. Their bold plans call for collaboration between stakeholder groups to create physical assets with finance and blockchains but without government, banks or corporations being involved. It is a company worth watching in the future, given its aim of creating the 'blockchain center of the universe'.

The government of The Netherlands has also been in the vanguard of leading on blockchain projects. It has been working with the UN, the World Bank and the European Union to develop a myriad of partnerships from releasing free books explaining blockchain technology, to toxic waste transport and authorisations in healthcare processes.[11]

BRE have cautiously welcomed blockchains, highlighting that it is early days for the technology but it should be closely monitored, given its potential to positively disrupt the construction industry. However, it rightly flags up the possible need for new governance and legislation to cover blockchains to safeguard its success.[12]

Digital learning points

Box 17.1 Digital learning points: blockchains

1   Blockchains are a digital ledger of transactions and a single source of truth.
2   Using blockchain technology, inspections could trigger payments to a supplier or subcontractor. Quality professionals need to appreciate the impact and potential pressure placed on them to accept or reject the work, given the new financial impact.
3   Blockchains can be used to reliably verify the origins of materials through the supply chain.
4   Used with sensors, blockchains can track how materials and construction components are handled through transportation and storage.
5   Blockchains can provide a new robust approach to auditing with digital signatures providing greater transparency and accountability.

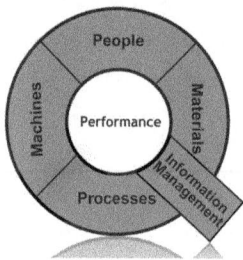

Notes

1   Construction Supply Chain Payment Charter. Retrieved from http://ppc.promptpaymentcode.org.uk/ppc/cscpc_signatory.a4d
2   Funding Options. Retrieved from www.fundingoptions.com/latest/
3   Department for Business, Innovation & Skills, *Construction 2025: Strategy*. 2, July 2013. Retrieved from www.gov.uk/government/publications/construction-2025-strategy
4   *Construction Manager*, 'Construction bodies call for an end to retentions'. 24 January 2018. Retrieved from www.constructionmanagermagazine.com/news/construction-bodies-call-end-retentions/
5   IBIS World, 'Title insurance in the US: US industry market research report'. September 2017. Retrieved from www.ibisworld.com/industry-trends/specialized-market-research-reports/advisory-financial-services/specialist-insurance-lines/title-insurance.html
6   Brickschain, 'Connect everything to everyone'. Retrieved from www.brickschain.com
7   buildingSMART, 'IFC overview summary'. Retrieved from www.buildingsmart-tech.org/specifications/ifc-overview
8   Buildcoin Foundation. Retrieved from www.buildcoinfoundation.org
9   Blockchains L.L.C. Retrieved from https://blockchains.com
10 News4, 'Blockchains L.L.C. proposes "Smart City" east of Reno'. 1 November 2018. Retrieved from https://mynews4.com/news/local/smart-city-to-be-build-at-tahoe-reno-industrial-center

11 Dutch Government. 'Blockchain projects'. Retrieved from www.blockchainpilots.nl/ results

12 BRE Group, 'Blockchain feasibility and opportunity assessment 2018'. January 2018. Retrieved from https://bregroup.com/wp-content/uploads/2018/02/99330-BRE-Briefing-Paper-blockchain-A4-20pp-WEB.pdf

18 Artificial Intelligence (AI)

I was at a BIM conference in 2015 in Dublin and one of the speakers held up his mobile and said, 'Well, we all have AI now.' To be honest, I didn't realise that by bringing up my photos on my iPhone I could put in a search term, press return, and it would show me all the results using artificial intelligence. Admittedly, the results for dogs missed a few photos of the collies and for some unknown reason came back with one photo of a bicycle but all in all it was right 90 per cent of the time in finding dogs from thousands of photos, without any manual tagging by me. That was when the penny dropped on the great potential for some really smart software to take construction into a new land of opportunity. I appreciate that some avid technology readers will have known about AI in phones for many years but there are large parts of the construction industry that would struggle to know that and appreciate the coming impact.

Artificial Intelligence is the academic theory and practical development of computer systems that can perform human-like tasks, such as speech recognition, visual understanding and making decisions.

Alan Turing was a brilliant English computer scientist and mathematician who, during the Second World War, was also a cryptanalyst, working for the British Government's codebreaking centre at Bletchley Park. Turing's critical role in Britain's codebreaking centre produced Ultra intelligence which broke the German ciphers used in their Enigma machines, ultimately shortening the war.

In 1950, he proposed the 'imitation game', what became known as the Turing Test, as a method to see if a computer could successfully imitate a human in providing answers to questions.[1] Turing suggested that the interrogator would be in a separate room to ask the questions through a teleprinter and the answers would be typed in reply. The test is whether the machine's replies may make the interrogator think it is a human through its emotional intelligence and natural responses, as opposed to intellectual intelligence. For decades, the computer scientific community pursued trying to mimic the human brain processes to develop AI with limited success. In 1990, Rodney Brooks published a paper entitled 'Elephants don't play chess', proposing a bottom-up approach to AI learning, using neural networks instead of the more fashionable top-down highly structured approach. By seeking out patterns in data, Brooks suggested that AI would make greater strides in development, which turned out to be correct. So, for example, when faced with deciding the number between 1 and 10 that has been handwritten as a squiggle, a neural network will use probability to assign a percentage to a drawing of the small loop towards the top, as being, say, either a 2, 3, 8 or 9. When it recognises another loop joined at the bottom it will calculate the probability as highest for an 8. However, before that happens, it has to likewise assign probabilities to each part of the loop to understand that it is a loop. To do that, it starts with the full range of options by taking the image broken down into a square of pixels. Then where the pixels are coloured differently to make up the edges of each number from 1 to 10, it assigns probabilities of likeliest options fed through neural layers until it assigns the highest probability to a specific number. A first layer is the input of all pixels on, say, a 28×28 grid, in which the number is written. The second layer may process edges based on prescribed weighted rules, the third layer may process patterns, and finally the fourth layer is the recognition of the digits between 1 and 10. There could be more or less layers, depending upon many design factors to solve the problem but this is the general concept.

To do all the calculations, it needs sets of examples to work from, given the variations in people's handwriting, and it will filter the probabilities through the layers before venturing on what it thinks it 'sees', i.e. an 8.

However, for the computer to learn by itself, it needs to be trained to know what is a 'right' answer. To teach the AI, it needs help to start with in knowing different handwritten 8s are 8s and not a very similar-looking figure, which is actually drawn as a 9. As the AI learns from the established training data using algorithms, it can then continue to learn by itself on new data as it computes the probabilities from more and more data examples. The handwritten characters that can be read are the central solution to machines reading addresses on envelopes that negates having to manually sort them during the postal process.

Algorithms are sets of rules for calculations, which in this case use weights and bias to ultimately decide on probability. Now it won't be perfect but the more complex the layers and calculations, and the more data it has to learn from, the better the accuracy, which may head to over 99 per cent correct results of whether a handwritten number is between 1 and 10. It is this learning from the data that is called Machine Learning as a subset of AI. Within Machine Learning is Deep Learning that uses new techniques for learning from unstructured or unlabelled data.

In a similar way, a computer neural network can learn to recognise a cat or a dog, as each image is broken down into millions of coloured pixels that make up edges and then patterns and finally shapes. Those whiskers belong to a dog and not a cat. That is a dog's tail and not a cat's tail, and so on. The concept can be applied to moving pictures of a person walking too near an excavator for the algorithms to calculate a risk factor, after learning from massive amounts of other video evidence. Likewise, audio recordings can be parsed into sounds which combine into syllables and hence into words and then sentences or simply phrases. The more data examples, the better the accuracy of learning the dialects and phrases around the world, from Glaswegians in Scotland to Northern Hausa dialect of Arewa in Nigeria.

So, AI is a glorified mathematics calculus and not some sci-fi, humanoid robot with a mind of its own trying to take over the world … yet.

Artificial Narrow Intelligence (ANI), also known as Weak AI, is the most successful type of AI, using algorithms and Deep Learning, so that the ANI can produce rapid stunningly impressive results, within a specific, clearly defined purpose.

Arthur Samuel from Kansas with a background at Bell Laboratories, who aided development of radar improvements in the Second World War, created a checkers computer game that helped uplift IBM shares overnight. He taught a computer to play the game and learn from its own performance, which, given the limitations of computers in 1959, was an astonishing accomplishment. Decades later, IBM's Deep Blue beat world chess champion Garry Kasparov in 1997 and four years later, Rodney Brook's own company, iRobot, developed the commercially available automated vacuum cleaner called Roomba. In 2008, Apple's iPhone included a speech recognition app for the first time that evolved into Siri with Google and Amazon quickly followed with respectively Assistant and Alexa, by using huge data patterns based on probability to learn.

IBM's Watson (Deep Blue's successor) computer system defeated legendary champions Brad Rutter and Ken Jennings in the US game show of *Jeopardy!* in 2011 and AlphaGo defeated eighteen-time world champion Lee Sedol in a five-game match of Go in 2016.

In 2018, Google Duplex[2] was showcased to make restaurant bookings over the telephone using an AI assistant that the *LA Times* suggested was 'nearly flawless' including pauses in conversations and ums and ahs. With downloadable apps and cheap software available to use AI each day, the ever expanding specific uses of ANI continue to grow.

However, Artificial General Intelligence (AGI), also known as Strong AI, is some way off, since creating a software that can mimic the workings of the human mind is fiendishly difficult. However, once achieved, it could increase its intelligence from learning by itself. The next level of AI is Artificial Super Intelligence (ASI), which, it is proposed, will be more advanced than the human mind. In addition, ASI may be reached incredibly quickly soon after AGI is achieved, since the increases in intelligence could occur at an exponential rate, called the Law of Accelerating Returns. The range in predictions of a few hours to days or weeks for AGI to leap to ASI leads to science fiction forecasts of utopian cures of cancer overnight and dire warnings from a range of eminent people such as Professor Stephen Hawking and Tesla's Elon Musk that it could spell the end of humanity. This leap, known as Singularity, has been predicted to happen in the mid-2020s to 2045 to never, depending upon various factors of AI capability.

For the short term, there can be little doubt that year by year, AI will impact upon business and continue to affect job types and the worldwide economy. The race has begun in earnest in 2018 to become the global AI leaders with the USA and China investing heavily and the UK and France not far behind. President

Macron announced that the French government will invest €1.5 billion in AI research over the four years to 2022.[3] In the UK, a new Government Office for Artificial Intelligence has been created to be advised by Dr Demis Hassabis, co-founder of DeepMind, and the Department for Digital, Culture, Media & Sport's new AI Council will be chaired by Tabitha Goldstaub, co-founder of CognitionX. The Industrial Strategy has also made AI and Big Data one of the four Grand Challenges in the years ahead (together with clean growth; the future of mobility; and meeting the needs of an ageing society) with the stated aim of the UK as 'We will put the UK at the forefront of the artificial intelligence and data revolution.'[4] A flurry of reports throughout 2018 included a mammoth and influential overview of the AI landscape and a directory of over 1,000 companies developing AI solutions, calling it the 'Cambrian Explosion of AI in the UK'.[5]

IBM have reported that, in manufacturing, 89 per cent of surveyed CxOs from outperforming industrial products companies say they plan to invest in AI/cognitive quality control.[6]

PriceWaterhouseCoopers have predicted that AI could contribute up to $15.7 trillion to the global economy by 2030, with a quarter of that value in improved quality, although it focuses on the automotive, financial services, technology, retail and communications sectors above energy and transport, without specifically mentioning construction.[7]

In fact, McKinsey found that construction was bottom of the heap in adopting AI by investment, over the past three years.[8] Some early uses of AI in construction have been on project management, by reviewing alternatives in delivery and proposing optimised recommendations. Other developments include image recognition to identify unsafe behaviour by site operatives, designing bespoke training and prioritising preventative building maintenance using data from sensors. Compared to other industries, these are modest test cases but whereas many other industries have large data sets for AI to learn from, construction has an enormous amount of information but rarely is it digitally formatted and accessible.

Construction is an industry made up of so many snippets of information from emails, water cooler conversations, telephone calls, weighty books of design calculations, and so on but often decisions are made with incomplete or plain wrong information, having costly or inefficient outcomes. AI can play a significant role in providing better information and even making recommendations to decision-makers, based upon previous training data. In a meeting that takes a key decision in design, having an AI set of options based upon many other projects facing a similar design problem will, over time, increase the probability that the options will minimise risk and increase success.

In Thailand, Ananda Development, the second largest property company in the country, has grown 1,000 per cent over the past seven years. It is leveraging technology in urban design, such as AI linked to driverless cars, IoT and robotics to create a different business model. Ananda has been hugely profitable from building skyscrapers near to rail stations. Usually people will pay a premium to live within 300 metres of a train station to commute to work. In the sweltering heat of Bangkok, walking further than that can be a real turn-off.

However, if they can take the lift downstairs to a waiting driverless, air-conditioned taxi that can efficiently navigate the streets to the office or factory then the demand for apartments in close proximity to stations may start to decrease and opens up opportunities for Ananda to build high rise apartments further away where land prices are lower. Such a shift in business strategy is based upon AI assessing the probabilities of successful outcomes in terms of geographic buyer purchases further away, using millions of data profiles of people and assessing when reliable, driverless cars will be available to implement a long-term investment. It is all about risk in asset management.

Ananda Development is investing in its staff to embrace the exciting technological changes to find new solutions in construction management. They see that staff who embrace new ideas and seek out how to use the technologies in their business will inoculate the business from staff who may naturally recoil against the sweeping disruption. By also offering digital services to property buyers, it increases the attractiveness of its properties, such as IoT within apartments using Big Data to understand their requirements.[9] Connecting these dots between technologies is how conversion will optimise efficiency gains in the future and allow accelerated learning by AI.

ClashMEP is generative design software that has been developed by Building System Planning, based in California, using machine learning to develop 3D models of mechanical, electrical, and plumbing systems.[10] It can avoid clashes with the architecture, such as automatically avoiding too many cables in a small ducting, by specifying cabling start and end points in the building design.

Autodesk's BIM 360 Project IQ uses connected data and machine learning to be able to predict project risks.[11] With data sources of video, audio, photographs and construction documents, Project IQ can make recommendations on risk mitigation measures. The software can identify level of risk in water penetration of a reinforced concrete slab and identify the highest risk subcontractors from levels of open issues, such as non-conformances and levels of re-work. The BI platform then can visualise the reporting data and allow drill downs so that project managers can get to the specific issues for a subcontractor to find out why they are deemed higher risk.

Smartvid.io is software that can lower safety risks through reviewing photographs and videos, using its learning technology, the Very Intelligent Neural Network for Insight & Evaluation (VINNIE), which uses a Deep Learning model to automatically tag the construction data and propose safety measures for the client.[12] From its training data it can query if someone on site is wearing the correct PPE or it can see that a ladder is not tied off for a supervisor to assess and perform corrective action, as required. The software is also very good at understanding what is in images, compared to traditional searches, which rely on manual labels attributed to them. Smartvid.io can recognise objects like a duct or a door and bring up those images on the search, providing a greater richness to search. When a project may have hundreds of thousands of photos and videos, then having intelligent search capability using AI can vastly improve the information flows around the project team and feed into workflows, such as a team talk the next day that picks up on those not wearing PPE.

Likewise, speech recognition will allow a quality engineer on a daily walkabout through the site to orally capture issues relating to photos that are being simultaneously taken and then translate the oral words into text for reporting or search, back in the office.

Volvo Construction Equipment has developed Density Direct for use on highway rollers that compact asphalt, which is a tablet with a preloaded app, an accelerometer, infrared mat temperature sensors on the front and rear, a GPS antenna and a base station with optional GPS rover.[13] In real time, as the roller passes over the asphalt, the density of the material can be read to quality assure the required material stiffness over the whole surface. Artificial neural networks are used to constantly learn as the roller vibrates over the material producing a density calculation. This produces better results than the traditional compaction meter value developed in the 1970s, which is a measure of compaction.

Academic work has shown that AI can be practically used in construction supplies of basic raw materials, such as aggregates.[14] Developing an artificial neural network (ANN) to quantify gradation properties can categorise aggregate stone into different types and sizes. As the aggregates move along a conveyor belt, a laser scanner creates 3D images of each stone and then a 2D 'wavelet' assesses the images. The AI can then identify the texture characteristics and allow each stone to be segregated into different classifications.

Compared to other industries, construction is virtually bottom of the league table in adopting AI. This is typical of the industry's conservative nature to all things technological. However, the potential is vast for AI in construction, with these as some examples:

- logistics planning for material deliveries;
- forecasting project risks;
- identifying defects in execution of construction;
- proactively identifying potential project disputes and 'talent pain points' in skilled labour shortages;
- turnover of staff;
- optimising the design of an organisation.[15]

The industry collects huge quantities of data, information and knowledge but is very inefficient in using it. AI can cut through the silos of data sources not only to track patterns and trends but also to start to offer suggestions and options that distil all that data and begins to make sense of it.

The industry has a duty, however, to collate the data, information and knowledge in better ways. How often do we have people with thirty or forty years of experience leaving the industry. and yet companies fail to 'download' all that knowledge and expertise? Exit interviews should not be a thirty-minute meeting, but a series of interviews to record the stories and problems that each person faced so that we can teach the AI about real-life cases. By offering a semi-retirement package over, say, 12 months and paying those subject matter experts to come back part-time, they can help pass on their knowledge in a structured way that can be recorded and not just passed on through training and mentoring. If companies then collaborated between those vast data banks of knowledge, we would save so much time (and money) from solving the same problems over and over, reinventing the wheel.

It intuitively feels right that we embrace AI in construction. As humans, we have created magnificent, stunning infrastructure but the scale of complexity grows with every new awe-inspiring design. We continue to find that executing a great design is so wasteful in time, money and resources. To dramatically achieve the efficiency that we crave in construction, AI can offer the solution. No single human being can maintain the requisite knowledge at all times through all phases of design, build and operation. On mega-projects, decades may pass and people will come and go. AI can be the constant. It can remove the inevitable silos that are currently created between individuals, inside joint ventures, within the supply chain and will be able to see through the chaos for a more scientific approach. We cannot possibly expect to efficiently build something with so many problems and decisions inherent in the design, construct and operate model.

Every day, thousands of individuals are making decisions based on incomplete knowledge and information. We are highly likely to see emergent AI systems among the Tier 1 contractors, with Bechtel AI, Costain AI, Balfour Beatty AI and Royal BAM AI being created, that will each assimilate their knowledge bases and evolve consciousnesses that fit with their own corporate values and vision. It is also likely that at least elements of those AI systems will collaborate to enrich joint venture solutions for clients.

Quality management has had a significant role to play in manufacturing problem-solving, through the dozens of quality tools, such as the eight disciplines (8D) model, FMEA (Failure Mode and Effects Analysis), TRIZ (Russian for Theory of Inventive Problem Solving), FTA (Fault Tree Analysis) and DoE (Design of Experiments). However, construction has rarely ventured past basic process mapping, Pareto and 5 Whys, and usually only in quality management workshops rather than through active usage in senior executive decision-making. We should encourage the use of more sophisticated tools that are commonly used in the automotive sector to solve advanced problems. Quality management AI is needed to push the frontiers of problem solving by taking the myriad of inputs to rationalise and process, with external data from long-range weather forecasts, commodity prices, staff capability records, plant telemetry, supply chain deliverability of materials and drone data (to name but a few), to provide a design manager and a construction director with insight into material selection for an architectural feature. Complicated? Absolutely. But think of how complicated it is for components on a new electric car or a nuclear reactor. That is why, as an industry, we need to embrace AI and shape and mould it, to solve difficult problems.

Challenges for AI

Challenges have already started to emerge. ISO 9001:2015 is based on principles that include 'leadership' and 'engagement of people' but how will this key international standard be applied when instead of humans, AI is carrying out the tasks? When AI is automatically taking decisions or instructing humans, then awkward issues arise for ISO 9001. Top management at construction companies may include AI at the executive level. A Hong Kong venture capital business, Deep Knowledge Ventures, has apparently appointed an algorithm named Vital (which stands for: Validating Investment Tool for Advancing Life Sciences) to its board of directors.[16] It has voting rights on investment issues and uses historical data-sets to uncover trends that may be missed by human analysts. Even if a human director remains in charge at a construction business, then AI could be appointed within the line management, providing advice to staff. The quality professional will need to appreciate how any AI quality management advice is being disseminated and used, in the same way it currently needs to understand the thread of top management's engagement through to the workforce. How will AI's role be organised and documented with reference to Clause 5.3 in the standard? Clause 7.1.2 calls out 'persons necessary' to implement the quality management system but if AI is used, then the standard does not cover such an eventuality. Clause 7.1.6 requires that an organisation should determine the 'knowledge necessary for the operation of its processes and to achieve conformity of products and services' and yet it could be AI that determines what quality management information is needed.

It has been suggested that between 20 and 30 per cent of design activities could be carried out by AI. Clause 8.3 on design becomes fraught with difficulties if lightning-fast 'black box' activities, involving decisions that cannot be unpicked, have been carried out by AI. Where is the audit trail for those decisions that may have changed the design?

Clause 9.3 on Management Review will sensibly need to understand that AI could be a critical input on reporting and may become responsible in implementing the outputs in terms of changes to processes. Or, are AI reviews needed to constantly assess performance which will be allowed to take the place of Management Reviews?

These are a handful of queries that start to percolate to the surface when trying to understand how to evaluate whether a construction business could be in compliance with ISO 9001, where AI is being used within business processes and its organisation. When AI is considered alongside ISO 9001, I would suggest that the next scheduled edition of the standard some time in 2020–2022 must assess how this critical technology (and other technological innovations discussed in this book) will need to be embraced, so that the standard remains relevant and useful.

The opportunities for AI to raise quality standards is very significant. AI may advise an operative on when and how to apply problem-solving tools, especially when they are complex, such as TRIZ. This could be an invaluable asset in more efficiently using such tools and may supply case studies or carry out calculations and provide answers on probability of the impact of inspection and testing results. Instead of human hold and witness points, AI will be on hand through cameras and sensors to provide a go/no go on moving to the next construction task, all the while recording evidence, every single step of the way.

The AI outputs to clients will include a BIM model, estimated capital expenditure (CAPEX), operating expenditure (OPEX) and operating performance,[17] with complete transparency from conception to handover, suggesting improved levels of customer satisfaction. A tricky challenge ahead will be, who owns the data involved throughout this process? A client will assume they own the intellectual property rights and the data but that may bar contractors from using standardisation processes and replication of pre-fabricated components in BIM model designs. A question that, no doubt, will make the industry lawyers rub their hands with glee, wanting to solve it.

A fundamental question for AI (and Machine Learning) is, how do we know if it is correct? Who checks the AI? As AI takes more and more automatic decisions, where are the checks and balances on this black box to know if it is forecasting the best options and undertaking the task without human intervention? The volatile mix of algorithms, past data and probability means that it is indecipherable at times as to how it arrived at its results.

Some have argued that in effect you cannot test the AI and have to learn to trust it, however, there are tests that the quality professional should be pushing the AI specialist to explain. First, unit tests assess if the flow of data appears logical through the AI model. It can identify exceptions and report back on quirks that should be investigated. Second, performance tests can be carried out in different ways, according to the model being used.

If the AI is predicting a value, perhaps the increased risk to freshly poured concrete when the weather temperature drops below a determined value, then a Mean Square Error (MSE) provides the average amount by which the model will be incorrect. The higher the MSE value, the worse the model has performed. Another such measure is the R-squared score and there may be false negatives and false positives from such tests and scores but the quality professional does not need to be an expert in such scores and mathematical systems. The point is to question the person responsible for creating or purchasing the AI and demanding answers on how it has been tested and to what standard. This does not happen at the present time and, given the IT industry's inherent defensiveness on all things technical, this could be an area of conflict, if not handled well. The quality professional needs to be assertive but polite in persistently querying if the AI has passed acceptable unit and performance tests rather than accepting vague promises of 'Yes, it's fine.'

In turn, quality management AI queries (and non-conformances) appearing in audit reports will help to galvanise IT and Information Systems leaders to demand answers from AI suppliers that can be translated into plain English replies. We need to be firm in the face of obfuscation but the cultural root cause is more to do with IT/IS leaders not wanting to face up to the fact that they themselves may not have a clue on whether the AI is producing correct results. With time and the creation of AI standards and tests, this defensiveness will pass but be prepared for the arguments ahead.

The ultimate test will be comparing predictions and/or actual decisions made by AI with the real-world outcomes. Circumstances can still change in the course of the process but by analysing what has gone right and what has not performed as expected, so the AI will learn and humans will be able to trust AI.

Conclusion

Global surveys have shown that investment in AI, when carefully project managed, reaps financial dividends for early adopters, with 83 per cent claiming moderate or substantial benefits.[18] The evolution of AI in project management suggests that step-by-step AI can develop from the integration of data sets and chat bot assistant through to intelligent advice on project scheduling and project risks towards greater project autonomy on decisions.[19] To initiate and rapidly expand AI in construction and become the standard bearers for the technology, quality professionals should be identifying problems that AI can help solve, developing costed project briefs and gaining support for testing out AI solutions on real-world construction sites. That requires us to reach out to the AI experts in academia and IT to bring together the right blend of subject matter experts to turn algorithms into added value construction.

Digital learning points

Box 18.1 Digital learning points: artificial intelligence

1   Appreciate the differences between different types of AI: Artificial Narrow Intelligence (ANI) = Weak AI, Artificial General Intelligence (AGI) = Strong AI, and Artificial Super Intelligence (ASI). We are nowhere near robots taking over the world!

2   AI used in construction is connecting data and machine learning to predict project risks (e.g. Autodesk Project IQ).

3   Risk management is aided by smartvid.io-type software that geo-tags features and then learns to identify quality management trends and patterns for defects.

4   The next edition of the ISO standard, ISO 9001:2025, will need to cross the AI bridge, so start to think about the impact on future certification.

5   AI's prediction and forecasting ability will rise rapidly in the coming years.

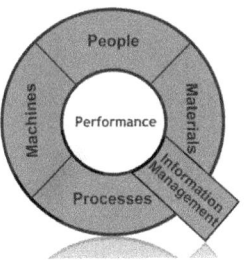

Notes

1 Turing, A.M., *Computing Machinery and Intelligence*. (1950). Retrieved from https://academic.oup.com/mind/article/LIX/236/433/986238

2 Leviathan, Y. and Matias, Y., 'Google Duplex: an AI system for accomplishing real-world tasks over the phone'. Google AI Blog. 8 May 2018. Retrieved from https://ai.googleblog.com/2018/05/duplex-ai-system-for-natural-conversation.html

3 Gershgorn, D., 'AI is the new space race: Here's what the biggest countries are doing'. *Quartz*. 2 May 2018. Retrieved from https://qz.com/1264673/ai-is-the-new-space-race-heres-what-the-biggest-countries-are-doing/

4 UK government, 'Industrial strategy: building a Britain fit for the future'. White Paper, p. 10. CM9529. (2017). Retrieved from https://assets.publishing.service.gov.uk/government/uploads/system/uploads/attachment_data/file/730043/industrial-strategy-white-paper-print-ready-a4-version.pdf

5 Big Innovation Centre/Deep Knowledge Analytics/APPG on AI, 'AI in UK – Artificial Intelligence Industry Landscape Q3/2018'. (2018). Retrieved from http://analytics.dkv.global/data/pdf/AI-in-UK/AI-in-UK-Full-Report.pdf

6 IBM, 'The artificial intelligence effect on industrial products'. Retrieved from https://public.dhe.ibm.com/common/ssi/ecm/17/en/17013217usen/industrial-products-ai_17013217USEN.pdf

7 PriceWaterhouseCoopers, 'Sizing the prize: What's the real value of AI for your business and how can you capitalise?' (2017). Retrieved from www.pwc.com/gx/en/issues/analytics/assets/pwc-ai-analysis-sizing-the-prize-report.pdf

8 McKinsey & Co, 'What AI can and can't do (yet) for your business'. (2018). Retrieved from www.mckinsey.com/business-functions/mckinsey-analytics/our-insights/what-ai-can-and-cant-do-yet-for-your-business

9 *The Nation*, 'Meeting the demands of posh condo buyers'. 9 July 2018. Retrieved from www.nationmultimedia.com/detail/Real_Estate/30349402

10 Building System Planning Inc., 'ClashMEP'. Retrieved from https://buildingsp.com/index.php/products/clashmep

11 Autodesk, 'BIM 360 Project IQ'. (2016). Retrieved from https://knowledge.autodesk.com/support/bim-360/learn-explore/caas/video/youtube/watch-v-nuWlpqevfIk.html

12 Smartvid.io, 'Tap into the power of AI (artificial intelligence) to reduce risk on your projects'. Retrieved from www.smartvid.io

13 Volvo CE, 'Compact assist for asphalt with Density Direct'. Retrieved from www.volvoce.com/united-states/en-us/products/other-products/density-direct/

14 Kim, H., Haas, C.T. and Rauch, A.F., 'AI based quality control of aggregate production'. (2003) Retrieved from www.semanticscholar.org/paper/Artificial-Intelligence-Based-Quality-Control-of-Kim-Haas/6c0d58e953b8546fb9f86223c6ebefd78df6d423

15 Blanco, J.L., Fuchs, S., Parsons, M. and Ribeirinho, M.J., 'Artificial intelligence: construction technology's next frontier'. McKinsey. (2018). Retrieved from www.mckinsey.com/industries/capital-projects-and-infrastructure/our-insights/artificial-intelligence-construction-technologys-next-frontier

16 *Business Insider*, 'A venture capital firm just named an algorithm to its board of directors – here's what it actually does'. 23 May 2014. Retrieved from www.businessinsider.com/vital-named-to-board-2014-5?IR=T

17 World Economic Forum, *Future Scenarios and Implications for the Industry*. (2018). Retrieved from www3.weforum.org/docs/Future_Scenarios_Implications_Industry_report_2018.pdf

18 Loucks, J., Davenport, T. and Schatsky, D., 'Early adopters combine bullish enthusiasm with strategic investments'. In *State of AI in the Enterprise*, 2nd edn. (Deloitte, 2018). Retrieved from www2.deloitte.com/insights/us/en/focus/cognitive-technologies/state-of-ai-and-intelligent-automation-in-business-survey.html

19 PriceWaterhouseCoopers, 'AI will transform project management. Are you ready? Transformation assurance'. (2018). Retrieved from https://news.pwc.ch/wp-content/uploads/2018/04/AI-will-transform-PM-Whitepaper_EN_web.pdf

19 Advanced materials science

Quality professionals have been leading on the quality assurance of construction materials since the 1980s, with the role of the Clerks of Works taking the lead in quality control of construction materials, throughout the centuries. With the contraction of the numbers of Clerks of Works on most construction sites, it has fallen on site managers with the support of quality professionals in recent times, to monitor and control the quality of the materials used.

From assessing the purchasing processes, transportation and storage of materials, through to certification of the materials themselves to demonstrate fitness for purpose, the quality profession needs to step up its responsibilities in checking that materials meet the required standards and specifications. The science of advanced materials is progressing rapidly, with new improved attributes for mechanical, physical, chemical and manufacturing properties.

Traditionally, quality control for construction materials on site included visual inspections and reviewing material tests. Unlike some of the hardware and software products we have discussed, such as drones and augmented reality that have already started to appear on construction sites, advanced materials are only just starting the journey towards commercial viability. Nevertheless, positive examples of alternatives to traditional materials are appearing.

Slump tests measure workability and consistency across wet concrete samples using a metal cone, tamping down the concrete using a metal rod and then measuring the slump, after the cone is removed. Concrete cubes are tested under a compressive machine at typically seven and twenty-eight days to measure strength. For in situ concrete, engineers can also use a Schmidt rebound hammer and the Windsor Probe test to determine the compressive strength. The challenge with these (and other) tests is that the results are invariably only a snapshot from the sample areas and hence batches of deficient concrete may significantly affect the results. Often these results are only available after the in situ concrete has begun to cure, making it very difficult to repair or replace, if found to be substandard.

While providing a full-time testing job, when large quantities of concrete are poured, these are antiquated methods of testing in the digital age.

Since the Second World War, it has been understood that there is a reliable relationship between strength gain to temperature in concrete. In warm weather, concrete gains its strength more quickly than in colder weather. By taking temperature readings in fresh concrete at set intervals, a meter can determine a 'maturity number', which allows a reliable estimation to be made of the concrete strength.

SmartRock wireless sensors are an innovative solution attached to the rebar, providing real-time monitoring through a smartphone app of concrete temperature, maturity and strength.[1] This type of sensor provides accurate results from precise areas within walls or on slabs and the results can last several years, allowing predictive maintenance to be carried out. They avoid the time delays of preparing samples for testing, transporting them to laboratories and waiting for the results to be reported. Other concrete testing techniques may not give reliable results for in situ locations, and there are fewer risks with sensors against the traditional numbering of samples and monitoring against approximate locations, which can introduce errors and mistakes.

For brickwork, a range of basic tests in both laboratory settings and on site include: compressive testing machines, water absorption and efflorescence tests by immersing in fresh water, and visual inspections for colour, size and shape. Certain types of Schmidt hammer can be used to test in situ strength. A digital approach can test samples using software such as WUFI (Wärme Und Feuchte Instationär,[2] translated as heat and moisture transiency) developed by the Fraunhofer Institute of Building Physics to measure the risk of surface condensation, mould growth and efflorescence on internal surfaces of walls and freezing on the external surfaces. Lucideon were used as the testing consultants on the Battersea Power Station project in London to assess if the brickwork, which had been built in 1927, was still suitable for use in the modern age after the building had been refurbished into residential accommodation[3] (it passed with flying colours).

In addition to using sensors and software to provide smart data on material performance, new, more advanced materials are being developed that make step-change differences.

Graphene is a single layer of carbon atoms arranged in a hexagonal framework that is the strongest material ever tested. It was first observed back in 1962 under an electron microscope and duly given the name by Hans-Peter Boehm, a German chemist. However, it was the Nobel Prize winners, Andre Geim and Konstantin Novoselov at the University of Manchester, who published game-changing research in 2004.

The University of Exeter has proposed making graphene concrete by including graphene in the mix to reduce by half the amount of mix materials required, to create viable concrete.[4] By doing so, there is a reduction of 446 kg/ton of carbon emissions. As a good conductor of electricity, proponents of graphene concrete have even discussed the possibility of using it in roads where it could melt snow and ice and even in walls as batteries to store electrical charge from solar panels.

A perennial problem for centuries has been concrete that spoils and cracks. Ghent University is researching into potential solutions of self-healing concrete.[5] One option is adding bacteria spores that react in contact with water that seeps into cracks. The bacteria precipitate calcium carbonate, which will fill in each crack. Other solutions include using hydrogels or elastic polymer materials inside capsules, added to the mix. All these experimental material solutions cost more than traditional concrete but, over the long term, save money from reduced maintenance. The same bacteria filling concrete cracks are also being tested as a soil binder by the University of Arkansas.

Ironically, when Vitruvius (in Chapter 2) discussed the potent effects of pozzolana concrete 'even when piers of it are constructed in the sea, they set hard under water', it took until modern times for scientists to understand why such concrete remained durable for 2,000 years.[6] Volcanic ash contains the mineral aluminium tobermorite, which, when it comes into contact with seawater, crystallises in the lime, as the curing process takes place, improving the compressive strength. With ongoing exposure to seawater, the aluminium tobermorite continued to grow over time, thus preventing cracks from appearing. The discovery has triggered opportunities for using the science in marine concrete structures today.

In Australia, fungus has been tested to bind together rice hulls with glass fines to create zero carbon, low-energy bricks for building. The bricks are fire-resistant and are less susceptible to termite damage. Given homeowners in Australia suffer from A$1.5 billion of termite infestations each year, this is not a novel material to be dismissed easily. The rice hulls and glass fines are waste products, further boosting the sustainability interest in developing this technology.[7]

The government in Vietnam has been encouraging the development of non-fired, ash and cinder bricks to replace clay bricks in construction and reduce the fumes from the kilns. Ambitious plans have not quite lived up to expectations of using up to 40 per cent of non-fired bricks by 2020 and clay-fired bricks remain the main building product, albeit the adobe (mixture of clay, sand and silt) bricks are slowly becoming more common. The challenge remains as the world has been waiting since 2007 for the Vietnam government to publish building standards approving the new bricks.

Charles McDonald, the Materials Engineer for the City of Phoenix in Arizona, developed a surface patching material using recycled tyre rubber.[8] For fifty years, construction has been recycling car tyres as an admixture into asphalt on roads, using a so-called 'wet process' with other innovative recycling also adding glass bottles and single-use plastics. This decreases waste products in landfill and improves the asphalt quality. In Australia, a senior engineering lecturer at RMIT University, Abbas Mohajerani has discovered cigarette butts can be added to asphalt,[9] which can still meet standards for heavy traffic use and improves heat conduction. In hot climates where cities typically suffer from the 'urban heat island (UHI) effect' from the absorption of the heat into the built environment. By innovating to add other products, such as cigarette butts, to the asphalt, improvements can be generated in dissipating heat faster.

The National University of Science and Technology, MISiS, in Moscow, has begun small-scale smelting of new grades of steel.[10] The vacuum induction melting furnace provides new types of alloys and improves ferrous and non-ferrous metallurgy.

Many of these new materials may not be commercially viable or will only find niche uses and, as with all new materials, there are significant regulatory hurdles to overcome before they are approved for widespread use. However, while it may take time for new materials to be regularly found on construction sites, quality professionals should be prepared for changes in the form of new and unfamiliar manufacturers' inspection and testing recommendations.

Nanotechnology is engineering at the molecular level, which either builds components (see Figure 19.1) from the bottom up by assembling through chemical molecular recognition or from the top down, using larger entities without the molecular control.

Nanotechnology exists on a range from 1 to 100 nm, where one nanometre is a billionth of one metre. If that is too difficult to imagine, then it is approximately the same ratio as a steel ball bearing to the Earth. Self-cleaning coatings have become the initial 'pin-ups' of the nano world that have grabbed media attention. Nanoparticles of titanium dioxide (TiO_2), aluminium oxide (Al_2O_3) or zinc oxide (ZnO) are coated on to construction ceramics or glass and clean the surface. TiO_2, for example, when exposed to UV light, eats into pollution or dirt before being washed away by the rain. Al_2O_3 is used to make surfaces scratch-resistant and ZnO can protect coatings from UV damage. All these nanoparticles help to protect against mould and fungus development.

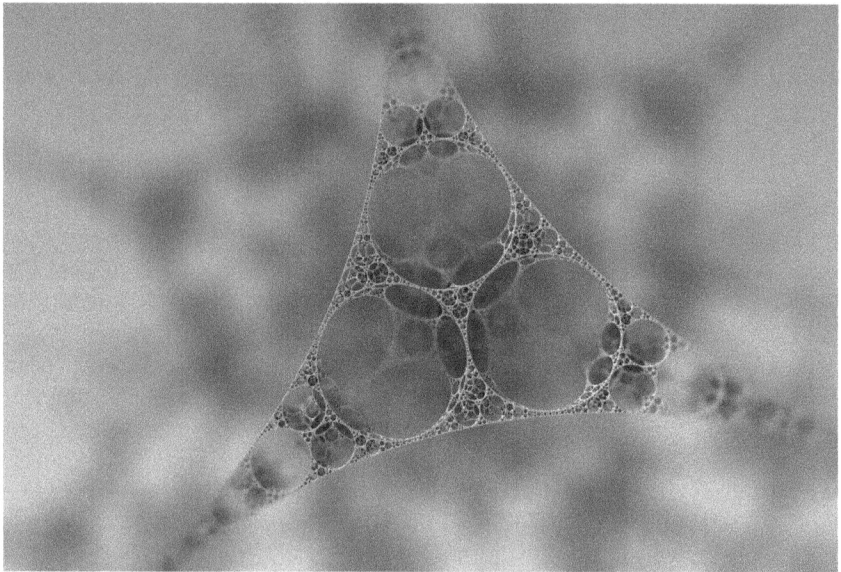

Figure 19.1 Nanotechnology that can produce tiny components.

Source: Pete Linforth, https://pixabay.com/en/users/TheDigitalArtist-202249/.

The thermal properties of carbon nanotubes can enhance ceramics and provide greater durability in concrete by preventing cracks from appearing. Likewise, nanosilica has been proven to strengthen concrete.[11]

There are many innovative architectural materials that can be used in design to capture certain effects. While they may be limited to improving aesthetic quality, the quality professional needs to understand how they have been tested and whether they meet required standards.

Litracon, produced in Hungary and developed by Hungarian architect Áron Losonczi, uses tiny optical fibres mixed into conventional concrete that create a translucent effect.[12] Liquid granite uses between 30 and 70 per cent recycled industrial base product, that is poured in the same way as regular concrete.[13] It has very high fire resistance and can be placed in hard-to-reach areas, to repair floors and walls.

Permeable or pervious concrete allows water to seep into the ground to mix with groundwater or feed into stormwater drainage.[14] With no sand but a highly adhesive paste to bind larger aggregates together, the water can flow through the porous material at the rate of $200 L/m^2/min$. It has a lower density and hence compressive strength but can still be produced for specific applications, such as pavements.

While 'transparent aluminium' was discussed in 1930s Germany, through to existing in the future in *Star Trek*, it is not metal at all, but a ceramic called aluminium oxynitride, comprising aluminium, oxygen, and nitrogen, known by the chemical formula AlON. It is incredibly tough and in tests 1.6 inches of AlON have performed better in ballistic tests firing a fifty calibre rifle round at it, than 3.7 inches of so-called bulletproof glass.[15] Practically speaking, while it is expensive in comparison to current materials, it has much higher resistance properties to scratches and breakages for windows.

With the increasing threat of climate change adversely affecting the planet, one nanotechnology borrows from plant technology, using the invention of a carbon-fixing polymer that can consume carbon into itself, with the potential for developing new protective coatings. Designed by MIT chemical engineers, led by Professor Michael Strano, post-doc Seon-Yeong Kwan, and others, the polymer has been proven to react with ambient light to absorb carbon dioxide from the air to strengthen and repair itself without the need for additive chemicals, mechanical stress or heat.[16] A next steps programme is being sponsored by the U.S. Department of Energy to develop it further.

Aerogels have extreme, dry material properties with a low density, solid framework of a gel. The most well-researched ones are the silica aerogels with a typical blue hue and sometimes shown in media photos above a Bunsen burner protecting a flower. They are 'mesoporous', nano-structured lightweight solids with the lowest thermal conductivity, meaning that they have between two and four times the insulation of traditional fibreglass or foam insulation. It is claimed that if aerogel with a density of $0.020 g cm^3$ was used to make Michelangelo's David, it would only weigh 2 kg.[17]

Advanced materials face challenges on cost grounds but that is because there is relatively little financial cost to the environmental degradation for open casting quarrying to take stone out of the ground or limestone and clay for cement production. Until there is a level playing field, with greater environmental taxes for extraction and pollution of using raw materials, then bringing advanced materials to market will remain difficult.

Pavegen has been around since 2009 and is a smart flooring solution that generates electricity from footsteps.[18] The operational data can establish hot spots of where users walk and can track footfall rates, including peak times, converting the energy gathered into on-demand lighting and advertising.

Solar Roadways use advanced technology and some recycled materials to create a superb solution to the current dumb road surfaces (see Figures 19.2 and 19.3). Hexagonal-shaped panels of tempered glass sit on top of solar panels that are laid in place of concrete or asphalt. Built-in heating elements can keep the road surface free from ice and snow, and LED lights can be used to create lines and signage on the surface that can adapt to different messaging for drivers with warnings of accidents.[19] The data that can be collected would be very useful in monitoring surface maintenance conditions, traffic, weather and accidents. Currently the products have been installed on driveways and car parks and have been tested in laboratories to meet US highway standards.

In Normandy, France, the world's first photovoltaic road surface has been constructed as a 1 km test phase, generating electricity for the village of Tourouvre-au-Perche. The asphalt surface has the ultra-thin, heavy-duty, skid-resistant photovoltaic panels made by WattWay[20] glued to the existing road surface to cover a 2,800 m^2 area for testing. Under the gloomy skies of northern

Figure 19.2 Solar Roadways® concept design for Sandpoint, Idaho.
Source: Solar Roadways®.

Figure 19.3 Solar Roadways® LED testing
Source: Solar Roadways®.

France the electricity generated has been moderate but the panels have survived well since their 2016 installation. The cost of Solar Roadways, Wattway and other similar products dramatically falls when production increases.

There has been some criticism that the design of such solar road surfaces will not solve all energy, safety and maintenance problems on highways but that is missing the point. As part of an overall solution to the myriad of environmental and construction issues we face, these products using innovative engineering of materials and technology, can help to solve the world's problems. Each product needs to be appropriately considered for local conditions and prioritised to meet the challenge of creating smart cities. They are not yet sustainable due to high R&D costs, but as with many technological products, unit production costs will fall as they evolve and demand increase. When smart roads meet self-driving vehicles, the opportunities will increase for safety and environmental solutions.

Digital learning points

Box 19.1 Digital learning points: advanced materials science

1   Wireless sensors and software can provide an alternative to traditional quality control testing of concrete and brickwork. Get to know how these data streams meet project specifications.

2   New advanced materials such as Graphene concrete, new grades of steel, nanotechnology, aerogels and carbon nanotubes will appear in specifications, and quality professionals need to understand how such specifications are being met. Are manufacturers' instructions being followed in transportation, handling, storage and installation?

3   Technologies, such as photovoltaic road surfaces and smart flooring and roadways, offer new functionality, and quality control testing at installation may be unorthodox or more sophisticated than traditional concrete and asphalt surfaces.

4   Smart cities will emerge with combinations of technologies that may have multiple impacts on quality controls. Using two technologies or advanced materials together may affect the stated performance compared to using them separately, requiring a risk assessment.

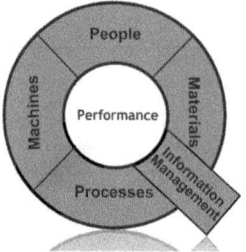

Notes

1   Giatec SmartRock. Retrieved from https://info.giatecscientific.com/smartrock-build-faster-safer-economically?campaignid=1018931886&adgroupid=60394335511&adid=2 94714198411&utm_campaign=SmartRock+Niche&utm_medium=search+ad&utm_ source=Google+Ads&utm_term=smartrock&utm_content=&gclid=CjwKCAjwjIHe BRAnEiwAhYT2hxd88CdQ2131fCX3MAFH2p6at2DnAmlk5nrlnYS7S6GBgOAOW v2jaRoC1VYQAvD_BwE

2   The Fraunhofer Institute for Building Physics WUFI. Retrieved from https://wufi.de/en/

3   Lucideon Limited, 'Brickwork evaluation of Battersea Power Station'. (2014). Retrieved from www.lucideon.com/construction/insight-hub/case-studies/brickwork-evaluation-of-battersea-power-station

4   Dimov, D., Amit, I., Gorrie, O., Barnes, M., Townsend, N., *et al.*, 'Ultrahigh performance nanoengineered Graphene–concrete composites for multifunctional applications'. (2018). Retrieved from https://onlinelibrary.wiley.com/doi/full/10.1002/ adfm.201705183

5   Snoeck, D., Van Tittelboom, K., Wang, J., Mignon, A., Feiteira, J. *et al.*, 'Self-healing of concrete'. Ghent University, 27 March 2014. Retrieved from www.ugent.be/ea/ structural-engineering/en/research/magnel/research/research3/selfhealing

6 *ArchDaily*, 'Scientists uncover the chemical secret behind Roman self-healing underwater concrete'. 5 July 2017. Retrieved from www.archdaily.com/875212/scientists-uncover-the-chemical-secret-behind-roman-self-healing-underwater-concrete

7 Huynh, T. and Jones, M., 'Scientists create new building material out of fungus, rice and glass'. (2018). Retrieved from https://phys.org/news/2018-06-scientists-material-fungus-rice-glass.html

8 U.S. Department of Transport. (2014). Retrieved from www.fhwa.dot.gov/pavement/pubs/hif14015.pdf

9 RMIT University, 'How brickmakers can help butt out litter'. 27 May 2016. Retrieved from www.rmit.edu.au/news/all-news/2016/may/how-brickmakers-can-help-butt-out-litter

10 MISIS, 'Metalloinvest and NUST MISIS to launch laboratory for development of new steel grades'. (2017). Retrieved from http://en.misis.ru/university/news/science/2018-01/5143/

11 Quercia, G. and Brouwers, H.J.H., 'Application of nano-silica (nS) in concrete mixtures'. (2010). Retrieved from www.researchgate.net/profile/George_Quercia_Bianchi/publication/257029738_Application_of_nano-silica_nS_in_concrete_mixtures/links/00b7d5243e5e804358000000.pdf

12 Litracon. Retrieved from www.litracon.hu/en

13 IMG. 'Liquid granite' Retrieved from www.img-limited.co.uk/product/liquid-granite/

14 National Ready Mixed Concrete Association. Retrieved from www.perviouspavement.org/index.html

15 Surmet's ALON®, 'Transparent Armor 50 caliber test'. (2011). Retrieved from www.youtube.com/watch?time_continue=4&v=RnUszxx2pYc

16 Chandler D., 'Self-healing material can build itself from carbon in the air'. Cambridge, MA: MIT. 11 October 2011. Retrieved from http://news.mit.edu/2018/self-healing-material-carbon-air-1011

17 Aerogel.org. Retrieved from www.aerogel.org/?p=3

18 Pavegen. Retrieved from www.pavegen.com/what-we-do

19 Solar Roadways. Retrieved from www.solarroadways.com/Home/Index

20 Colas Group, WattWay, 'Wattway, the Colas Solar Road'. 10 November 2016. Retrieved from www.youtube.com/watch?time_continue=2&v=OI9fSnBig3s

20 Future technology

Before we get to the human-free construction site and design office, we will take a breathtaking journey of experimentation and some technology trends that become dead ends, raising digital capabilities of humans handling the technology and sci-fi-like hardware and software that we have yet to even imagine.

How technology will morph and evolve is very difficult to attempt to forecast but there are seven clear themes emerging for quality professionals to understand and learn how to best adapt to, in order to optimise quality management and business results:

1. Digital fluency
2. People skills
3. Data explosion
4. Critical thinking
5. Thinking machines
6. Advanced materials
7. Artificial intelligence (AI)

1 Digital fluency

Accepting the fast-paced changes and constantly learning about new technology will mean the quality professional maintains confidence in a reasonable level of understanding of what is going on around them. That requires regularly reading technology blogs, playing with new apps on your smart phone, talking to younger people (digital natives) on what technology they use, buying or experimenting with gadgets, adopting a curious mind while remaining realistic that we need those bricks and mortar built within cost and to achieve customer satisfaction. This will stand the quality professional in good stead.

There are wider societal effects of change that touch upon technology and the quality professional needs to be cognisant of them. Long-term, permanent jobs will become rarer and more work will be fixed contracts with greater pay, based upon performance outcomes but working freelance. Quality professionals are likely to have a small portfolio of contracts, working longer, to seventy or seventy-five years old, as technological change negates the need for permanent staff. The performance outcomes of work will be automatically measured by technology and there will be fewer hiding places for people to coast through days. Understanding this technology means that the quality professional will need to raise their game and their value to their employing businesses.

With various reports indicating that jobs will be impacted over the coming years, Cisco and Oxford Economics predict US jobs will be hit: '4.3 million workers will be displaced (3 per cent of the total workforce), plus an additional 2.2 million workers will be disrupted resulting in a total of workforce 6.5 million job moves by 2027.'[1]

For construction jobs, the outlook is a 11 per cent fall in percentage terms but a positive 30 per cent increase in overall construction industry value, suggesting that there will be big investment but sustaining fewer jobs. Higher skilled jobs will see up to 6 per cent growth over those ten years. The message for quality professionals is to upskill as a priority, especially with digital capabilities.

In construction, BIM digital modelling has created the opportunity for the eighth dimension to become Quality Management. After centuries of 2D paper, BIM is virtual 3D, time (scheduling) is 4D and cost (programming) is 5D. There have been proposals for project lifecycle information[2] to become 6D, although others in the industry[3] have suggested sustainability for 6D, and facilities management as 7D. Regardless of the numbering of the dimensions, a layer of quality management data to demonstrate identified defects and re-work, a quality score for data quality, materials performance data and other quality information, will demand expertise and ownership by quality professionals.

Such digital skills were highlighted in a viral video by Scott MacLeod and Karl Fisch called *Shift Happens*, in 2010 that grabbed headlines with the suggestion that 65 per cent of children entering primary school in that year would ultimately end up working in completely new job types that did not yet exist.[4]

Digital fluency entails seeing the world through the lens of data and information throughout the matter we see around us and the world we imagine. It is seeing all aspects of construction and quality in terms of what information we need to create and operate the physical construction. We need to break the mould of only using the staccato tools of written management systems, audits and inspection and testing as the only way to understand how we can quality control and quality assure the outcomes. We need to demand the information for all aspects and from all angles of people, processes, machines and materials to decipher and evidence the performance quality.

Our ubiquitous smartphones will be replaced by wearable technologies, especially smart glasses and new skills will be needed, such as eye staring to open and close apps on a head up display inside the glasses. They will allow both augmented and virtual realities to appear that will enhance our understanding of construction.

2 People skills

Appendix 1 provides a generic quality professional role profile for 2030. This will, of course, vary according to the seniority of the role but gives an idea as to the new skills, knowledge and capability needed in our profession in the future. Quality capability needs to evolve and adapt, so we must take a new look at people capability for construction work to assess if operatives have the knowledge, skills and expertise to deliver the required levels of quality in construction.

Taking even more time to understand other skill sets in IT, information systems, finance, design, construction and HR will improve the understanding of the complexities and process flows and open up the opportunities to identify improvements that add value to the business. It also bolsters the interpersonal skills, as we need to be more gregarious, outgoing and inquisitive. We need to network like never before, to build contacts and connect with a wide spread of people inside and outside of construction, so that there are bonds of trust and confidence between us and other industries. We can then use that trust to access useful information and ensure we can get under the surface of process maps and deeply understand what is going on inside the business. That creates opportunities to identify continual improvements through best practices, which in turn ignite innovation.

3 Data explosion

The amount of data is continuing to increase exponentially, making it impossible for quality professionals to manually manage and manipulate it all. It is becoming increasingly obvious that even modest amounts of data are not being fully understood and used. The answer is to use AI to collate and filter, to either provide options to the quality professional to act upon or for AI to take automated decisions within given parameters. The BI dashboard will become more predictive and indicative of the future. Current clumsy snapshot charts of past data and information will become nothing more than a pleasant, wistful pastime, like looking at footballing statistics and league tables from decades ago. AI data mining will be proactive, searching out every source of data within a business, without being told where to look in spreadsheets, databases, individual laptops and obscure software packages. Algorithms will identify and link to internal and external sources in order to provide the best insights in real time.

The quality professional will need to be agile in reading these predictions and forecasts and making recommendations to executive decision-makers. That places them closer to decision-making but also in the frontline of a fast-moving construction business. If they (and their AI) get it seriously wrong, then the importance of quality management will continue to diminish. However, if they get it right, then eyes will be fixed on them as being a key future component of strategic thinking. That will lead to enlightened executive boards creating a Chief Quality Officer (CQO) as a top professional within the business C-suite who will have significant authority as the live BI insights demonstrate the advantage of putting a quality professional at the heart of decision-making.

Data storage is likely to become much cheaper to the point it will be free and unlimited. Every machine, piece of equipment and many physical objects will be fitted with sensors to stream data. Clothing will become smarter and be fitted to the Internet. By 2030, infrastructure will have sensors buried in walls, floors and ceilings that will allow predictive maintenance but requiring them to be tested as they are built. Such tests will be carried out seamlessly by connecting to AI that reports to the quality professional through smart glasses and alerts them to any failed sensors before they are embedded into fresh, self-healing concrete. The data explosion will overwhelm humans, without such AI, and, as such, the quality professional needs to learn how to test AI to minimise the risk of errors in the algorithms and applications that help to manipulate and prioritise the data.

Data quality is a pressing concern today with typically a failure of ownership to embed quality assurance of data early into processes and test its accuracy, consistency, completeness, integrity and timeliness. Quality management should take ownership and link up between the IT and software professionals and data users to be the guardians of data quality assurance.

The explosion of data will also allow us to recreate construction sites, if we need to look back in time at how the infrastructure looked. If there is a query or an investigation, then using photogrammetry, we can piece together a moment in time for, say, a reinforced concrete cantilever before it disappeared behind a floor covering. The BIM digital model, together with laser scans, photos and videos taken during construction, would allow a 3D replica of what it looked like when it was being built. Software can search and find photos, even if they are not formally tagged to recreate the three-dimensional images, which can even be viewed in virtual reality to give more texture and detail. For an important root cause investigation, such imaging can offer important clues that may have been missed at the time and can be used in reports or for training videos at a later date.

4 Critical thinking

Critical thinking, by assessing the evidence from BI, will allow quality professionals to offer more insightful judgements that make a difference in how we as a profession can add better value. Critical thinkers have an ability to think rationally and logically to connect ideas and rigorously test hypotheses, before coming to a conclusion or recommendation.

Seeing across the traditional business departments, between different stakeholders, benchmarking against other industries and understanding the impact of many outside factors, will be key to the quality professional. We have traditionally been better than many at connecting the dots but in the future this capability will become even more important. Quality management has offered tools and methods to solve problems for decades and the profession should find critical thinking a natural skill to integrate into problem solving.

Working closely with our cousins in environmental management, health, safety and wellness, information security and business continuity will continue to grow as the common theme of lowering risk imposes a need to understand the overlaps and duplications between them.

AI will assist the critical thinking processes to summarise ideas and challenges. It will help choose the appropriate problem solving tools for a situation and even carry out calculations and assessments where those tools can currently be a challenge due to their complexity, such as TRIZ and Lean Six Sigma. This will improve the value of quality management to construction. However, learning the core theories behind these problem-solving tools will mean quality professionals are placed in a better position on advising how AI can practically assist the workplace of designers and construction experts. Given the newfound influence of the quality profession, as it seizes AI to open up access to the suite of problem-solving tools, so the risks increase that a quality professional will make a mistake and it will impact on the construction process. But if AI is harnessed, then who is responsible: the human or the AI? It will mean quality professionals need to enquire about professional indemnity insurance that covers them when AI is being used.

Critical thinking means seeing the landscape of knowledge and become protagonists in pushing knowledge management systems that codify business knowledge and structure it. The challenge is that knowledge workers can spend up to 36 per cent of their average day looking for the right information.[5] From these knowledge networks, quality professionals can have a better bird's-eye view of the information available and take a mature look at designing systems with AI that facilitate getting the right information to the right person at the right time.

Quality professionals should play a positive role in Communities of Practice and Communities of Experience to enrich the exchange of tacit information and knowledge and encourage the codification of it so others can find it.

5 Thinking machines

As the likes of Amazon join the construction industry to build their vast distribution centres,[6] they will bring with them robots, drones, autonomous vehicles, AR and all the other technology onto site. They will have no time for chaotic planning and manual labour, as their instinct will be to use the latest tools to solve age-old problems. These new entrants will help drag the rest of the industry along into scientific planning and programming of design and construction processes.

3D printers will soon start to appear on construction sites, at first, printing small components. As the speed quickens in printing and prices fall, so for simple repeatable components, such as door and window handles, signs, taps and letterbox frames, the 3D printer will sit inside the material storage yard, manufacturing large numbers, helping to reduce transportation and carbon emissions. Over time, the number of robots tying rebar and laying bricks, and

3D printers piping concrete into walls, will become common. As adoption grows, so the functionality of such machines will also grow. 3D-printed pre-fabrication may appear as much on the construction site as it will in the factory.

In the same way that we would ask for training and educational records relating to technical skills of those on site, so we will need to ask for and understand the certificates demonstrating the 'fit for purpose' programming and build quality of robots to allow the quality professional to perform an audit.

When I first saw a Total Station used on a construction site, I didn't need to know how the station was collecting data to be downloaded later in the site cabin, but I wanted to understand if it had been calibrated and if records existed to provide me with a level of confidence that what was being reported was reliable. I was no expert but by asking the question of the surveyor, I could probe deep enough to see if they could demonstrate confidence in the machine.

Humanoid robots will emerge in business in the 2020s and by the 2030s will appear on construction sites lifting, moving and fitting pre-fabricated materials. They will become as common as autonomous excavators, loaders, trucks and cranes on sites, resulting in a dramatic drop in the number of operatives on site and, as a result, improving safety.

Flying cars are still left for sci-fi films and prototypes but the number of drones in construction are growing rapidly, year on year, and their functionality will increase, as the list lengthens of construction tasks they can undertake, by flying faster, further and for longer. Quality professionals need to understand this technology now, not tomorrow, in order to enhance how they carry out on-site inspections and record images and data on those inspections. This will give us a greater, hands-on appreciation how other disciplines are using drones.

6 Advanced materials

Not only will sensors be embedded into concrete but sensitive meters with software will assess brickwork. This digitising of existing material performance, however, will not have as great effect as new advanced materials appearing in laboratories. These will improve the material properties of strength, endurance, plasticity, thermal conductivity and hardness in graphene, aerogels and carbon-fixing polymers.

Ensuring that these new dynamic properties, some of which are at a nano level, are fully understood and can be explained in simple terms, will be part of the quality professionals' remit. So will examining manufacturers' instructions and checking the authenticity of inspection and testing certificates.

Combining technologies in construction materials through solar paving and roads will push the boundaries of traditional quality management. How do we untangle the impact of using different materials in different environments on the operating performance? 4D printing of materials that change when exposed to temperature or moisture requires more sophisticated methods of quality control and a greater reliance on laboratory or third party certifications.

7 Artificial intelligence (AI)

The humanoid AI is a long way off yet (Figure 20.1), but more mundane AI software is with us now. To imagine the potential of basic artificial intelligence, take anything (X) and add AI. 'X + AI' will trigger a massive groundswell of start-ups and new technologies. Harnessing the opportunities of AI reporting from embedded sensors around the site and then shaping AI decisions for quality management tasks will place us in the vanguard of AI capability. The AI automatically sends out a written quality alert on material defects found in steel beam quality, even though the UK organisation has not determined the communication is required; a black box of AI has determined it, from the results in a factory of a steel supplier in China.

While I assume that ISO 9001 will be updated in a future edition to adapt to such scenarios, the quality professional needs to be on the look-out now for AI appearing, given the pace of change and the time delay in international standards sometimes reflecting what is happening on the ground.

Executives may be able to easily access or be provided with AI performance statistics on the accuracy of predictions from audit scores vs real world on-site results. Putting a quality professional's name to a report can raise the stakes, as AI will score the report's accuracy.

These seven themes will surge forward in the coming years, requiring a greater preparation by the quality profession than is indicated by many contracting

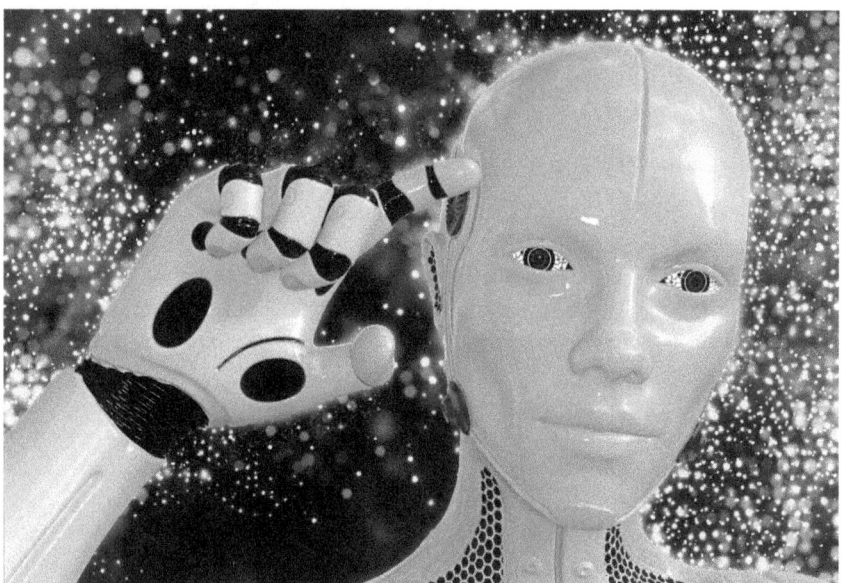

Figure 20.1 The humanoid AI may take decades, if ever, to appear but more mundane AI software can assist quality management today.

Source: Pete Linforth, https://pixabay.com/en/robotcyborgandroidrobotics-3737276/.

companies. Our response to the technological change needs to place the fundamental requirement of information management at the heart of our approach and using a quality assurance graded approach to understanding the level of risk can help us make informed decisions.

We need to step up to the technological challenges or face extinction.

Notes

1 Cisco & Oxford Economics, 'Modeling to inform the future of work'. (2017). Retrieved from www.cisco.com/c/dam/assets/csr/pdf/modeling-inform-future-work.pdf
2 NBS, 'BIM dimensions – 3D, 4D, 5D, 6D BIM explained'. (2017). Retrieved from www.thenbs.com/knowledge/bim-dimensions-3d-4d-5d-6d-bim-explained
3 Zhu, H., 'BIM: A "model" method'. The BIM hub. 17 October 2015. Retrieved from https://thebimhub.com/2015/10/17/bim-a-model-method/#.W8R6Oy_MxQI
4 Fisch, K., *Shift Happens* video. 12 July 2010. Retrieved from www.youtube.com/watch?time_continue=21&v=SBwT_09boxE
5 Schubmehl, D., 'Unlocking the hidden value of information'. IDC Community. 14 July 2014. Retrieved from https://idc-community.com/groups/it_agenda/bigdataanalytics/unlocking_the_hidden_value_of_information
6 Wilmore, J., 'Will Amazon enter the construction jungle?' *Construction News.* 8 May 2018. Retrieved from www.constructionnews.co.uk/analysis/cn-briefing/will-amazon-enter-the-construction-jungle/10030732.article

21 Road map to digital quality management

Appendix 2 describes a hypothetical scene in 2030 based on my interpretation of where we are and where we seem to be heading. Time, as always, will tell whether construction adopts and exploits the technology, in the way I've described.

So, how do you get from where, as a quality professional, you are now, to where you need to get to? There will be many variables from the size and financial health of the business, its digital appetite, the leadership and vision of both the CEO and CIO (or equivalents) and the quality professional's own personal values, beliefs and requirements for driving ahead with digital quality management.

However, I am going to sidestep most of the obstacles and concentrate on a way forward with a fair headwind. Whether that is reasonable, unlikely or stark staring mad, the reader can judge but others can argue for what they think *might* happen. I am arguing what I think *should* happen.

What follows are stepping stones towards the goal of full Digital Quality Management. For reference, Appendix 4 includes a summary of all the previous digital learning points from Chapters 6–19. The quality professional may only wish to move down the road of one or two steps or may appreciate the difficulties of the journey and take it stage by stage and assess and weigh up challenges and achievements before moving towards the next stepping stone. As ever, assessing risks is the key to making decisions and remember the journey never ends! Good luck!

Step 1 Leadership

> [E]very successful quality revolution has included the participation of upper management. We know of no exceptions.
>
> (Joseph Juran[1])

While progress can be made from middle manager levels, it will be very tough, if not highly unlikely. Not least, the problems of being buffeted by competing priorities from above and the failure of peers, one's work colleagues, to see Digital Quality Management as something they should take seriously.

The boss needs to be onboard – that means the CEO. Unfortunately, CEOs have the attention span of the speed of a return of a Google search engine result. Even the most digitally savvy and enthusiastic CEO has to prioritise banks and shareholders, lawsuits, financial audit reports, PR disasters, and so on. The solution is finding a champion on the Board, who will build alliances among other directors and senior managers, cajole and pester suspicious directors, smuggle digital objectives into the annual business plan and/or corner the CEO with a 10-second soundbite of devastatingly simple math to prove Digital Quality Management could double profits (it won't yet – so this is a desperate, last resort tactic and not recommended).

Once there is a buy-in, then there is leverage in resourcing a larger budget, more staff and better enabling technology. Don't tell CEOs, but they usually won't make time to take a deep interest in projects or business plans, unless something awful goes wrong. Once the leadership is secured in some sort of shape and form with a clear business objective or approved project, then the quality professional is up and running.

One tip: make sure that in your budget is an item for 'assessing the budget spend to confirm if it is delivering the required outcomes'. It is too easy to spend money without taking stock of whether we are actually achieving what we need to achieve. That 'taking stock' may also require funding itself, if it is to be done properly.

Step 2 The business objectives and planning

> Good things only happen when planned, bad things happen on their own.
> (Philip Crosby[2])

The business objectives need to spell out the commitment to an information management strategy, as part of a wider progressive strategy that embraces organisational development (OD) to develop people capability. From those vital few business objectives, the quality objectives need to set out how the required performance of construction services will be attained through quality management deliverables.

A business objective may be 'to create Level of Detail 300 for digital models for all construction projects' and one quality objective may be 'to assure the quality of 95 per cent of specified information impacting upon the performance of the building within each digital model'. Such measures are written from an agreed business baseline with the aim of moving towards an overall goal. If digital maturity is high and the business has been exhibiting year-on-year improvements, then the information quality may move beyond 95 per cent but if the business is just starting the digital journey with a Level 1 BIM, then it may choose 70 per cent of information quality. Only the business will know what is reasonable to target but it must choose some sort of target, given the crucial importance of information management.

A plan, any business plan, needs to be short. No one other than the author of the plan, one or two nerdy colleagues and the author's line manager will read the damn thing unless it is short, so spending weeks and weeks writing dozens of pages and mastering the intricacies of clipart in Word, is futile. It may look incredibly professional but no busy decision-maker will read and digest it, in most modern-day businesses. There isn't the time.

The plan needs to pithily spell out the problem, the solution, the cost and the outcomes in improved savings (whether in money and/or time) and quality. If detail has been asked for, then create a batch of appendices and stick the details of spreadsheets, diagrams and endless chatter in there, but otherwise on one page spell it out, so that anyone (and I mean anyone) can grasp what is being asked for and what they will get in return.

Step 3 Voice of the customer

> All of management's efforts for Kaizen boil down to two words: customer satisfaction.
>
> (Masaaki Imai[3])

Yep, that old chestnut is still right; find out what the customer thinks. No matter how good the product or service, there will still be complaints and ideas to improve. Get in front of the customers face to face by sweet talking them into a precious half hour (once they start talking the chances are they will end up giving you an hour or more, as everyone likes being asked for their opinions).

Understanding the final external customer's requirements provides a means to validate the verification of the design. But crucially understanding internal customer's requirements is just as important. The virtuous chain stretching from idea to operating the built environment will invariably pass through the con-tractors' internal processes that link internal customer to internal supplier repeatedly, until delivery to the external customer. Along the way it is likely that there will be numerous interactions with the external customer, such as project reviews and witness hold points.

From a quality standpoint, find out where you are heading as a final destina-tion. The customer may not be able to describe in explicit detail the numerous data and information (they may) but they should give a clear sense of what cus-tomer satisfaction looks like, in order to figure out the solution.

Step 4 Quality data and information

> Organizations, as information processing systems, will tend to produce invalid information for the important, risky, threatening issues (where ironically they need valid information badly) and valid information for the unimportant, routine issues.
>
> (Chris Argyris[4])

Digital Quality Management, as opposed to digitisation of quality management, will only happen if information management is at the core of project planning and its importance managed throughout the lifecycle of the project.

That means taking the project requirements and identifying all the information needed to demonstrate compliance with the following:

- the business end-to-end processes, including project processes;
- the contract specification, works information and material and process standards (depending upon the contract, there may be performance specifications based upon outcomes required, prescriptive specifications setting out particular materials and processes or even specifications for specific branded products and services that must be used);
- digital design models that allow 2D, 3D and 4D (time) representations

There needs to be comprehensive quality management of graded information and not just the traditional construction quality assurance, which will typically be limited to writing a PQP/DQP and having ISO 9001-aligned procedures for design change control, material standards, management system audits, document control, etc.

Rather, wherever information and data are needed in the project, there should be a graded quality management approach to, say, check inspection and testing routines to capture proof of meeting the performance requirements.

Information is typically missing, fragmented and unstructured from the designer, contractor and throughout the supply chain. A comprehensive, thorough approach to information management needs to ensure that prioritised information is created and managed where it is needed most, even if that's a Tier 4 subcontractor making small components, which could be critical to performance or safety.

The information and data captured initially may have to fit the techniques available at the time. Hence, aerial data from drones may be easier to harness as they are seen as a flavour of the month and are relatively cheap, than commissioning a purpose-built laser-scanning robot for a particularly inaccessible part of a road bridge that may arguably capture more useful data. However, the point is to map out the information required and identify the cost-effective techniques available to efficiently absorb the data.

Appendix 3 sets out the Quality Information approach to a simple construction for a reinforced concrete wall to prompt deeper thinking about the information requirements. Instead of blissfully assuming that the bid process and brief commercial checks will have assured the quality of the chosen subcontractor in every aspect, quality professionals need to be more robust in ferreting out the evidence to show that the performance can be achieved for the work required, *before* it starts.

Over time the use of AI to 'read' the documents and provide assessments and/or follow-up questions for audits will provide greater scrutiny of the evidence of Quality Information conformance.

Until Digital Quality Management becomes a tried and tested approach with data mountains available, then, in the early days, we may have to settle for second best that moves us closer to proven performance, even if it is not perfect.

Step 5 Intelligent analysis

> It is a capital mistake to theorise before one has data. Insensibly one begins to twist facts to suit theories, instead of theories to suit facts.
>
> (Sir Arthur Conan Doyle[5])

Data sources for quality, sensors, plant telemetry and RFI tags on materials, etc. should be experimented with, to determine effectiveness and build an understanding of the 'art of the possible' with BI.

With data and information collated, the analysis is a key step that is all too easy to get wrong. With the risk of 'data constipation', the wave of information bombarding the quality professional can be daunting and currently it is highly likely that mistakes will be made. Trying to sort, filter and sift the data and information results for patterns and trends can be difficult and can lead to false assumptions and up blind alleys. It is important that any conclusions and recommendations to executives come with a caveat and health warnings.

I faced the wrath of a director after spending months researching and developing a plan of action for taking BIM to Level 2, only to make an innocent assumption over the common data environment (CDE) costing, without such a caveat. The director regurgitated my plan to the CEO (probably within a few minutes) and the CDE cost was apparently under-reported. I had already solved the funding issue through agreement with regional directors but my costed plan wasn't clear.

Given such fears then, the simplest way is to have a second look (and third and fourth looks) by colleagues you can trust and who will be both insightful and truthful. You need honest feedback to reflect on which patterns and trends you think are being revealed by the data and information. Be prepared for some brusque, hard-hitting replies but such feedback is essential. Regardless of how many humans come up with suggestions and proposals for summarising data and information on a BI dashboard, to reduce the risk of bias and strengthen reporting of meaningful trends and patterns of quality performance, AI should be used, when possible.

Partnering with experts in AI to identify how the power of algorithms can be used to report on prioritised elements of the project can unearth significant information that may have been missed by only using the human eye. At the start, be patient. It takes time to understand what the problems are that require solving for AI experts to create the algorithms that will unearth useful analysis to be highlighted within a BI dashboard.

Step 6 Reporting

> First, when given a choice, 80% of people preferred sentences written in clear English and the more complex the issue, the greater that preference. But second, it found that the more educated the person, the more specialist their knowledge, the greater their preference for plain English.
>
> (Christopher Trudeau, Thomas M. Cooley Law School, Michigan[6])

For any narrative around data and information, ensure that it is simple and easy to understand. For English speakers, then, that's PLAIN English.

Ensure that digital information and data are available at all points of a project, so that live business intelligence reporting can be enabled into a dashboard that informs better decision-making, whether they are a client, designer or contractor representative. That way a single source of performance information is available to all stakeholders, building confidence and improving the likelihood of consensus over actual performance.

Usually the dashboard is created based upon specified reports that may or may not relate back to project objectives and KPIs. However, this can quickly become a fool's game of chasing metrics that are easy to source from simple databases, information that is months out of date or to data fit individual bias. However, very pretty pie charts, line graphs, tables of figures can colourfully decorate board reports but utterly fail to provide clear evidence of the quality of outcomes being achieved (see epigraph by Chris Argyris under Step 4). Staying focused on prioritised performance outcomes is essential for the dashboard to add value.

Hence, the quality professional should aim to influence business objectives and ensure they develop quality objectives, KPIs and metrics from them to add value to the business. By learning and understanding BI software solutions – Tableau, Microsoft BI, Qlik, Salesforce and Birst – it is easier to explain to business intelligence/IT teams the quality requirements.

Over the medium to long term, quality professionals should proactively aim to create and collaborate with the formation of Data Trusts to access greater insight from construction Big Data.

Oh, and remember, one page for executive reports (yes, I know, it's 'impossible').

Step 7 Learning and improving

> Learning without thought is labor lost; thought without learning is perilous.
>
> (Confucius[7])

Will we make mistakes? Yes. Is that a problem? No, not really. This is a new approach to quality management and if you want minimal risk, then stick to the tried and tested, little added value of the traditional approach to construction quality assurance. You can be incredibly busy and produce a variety of reports that may look impressive but fundamentally are not delivering what is needed for construction quality management. Scratching the surface with manual quality audits and trawling through a small number of manually reported nonconformances will not significantly reduce risk to the objective of performance outcomes of the built environment.

The road will be bumpy, as there will be defensiveness, arrogance, apathy, misunderstandings and ignorance but stay strong and resolute. Learn quickly and be confident of the DQM approach. Construction is decades behind other industries in its processes, people capability, machines and materials and, most of all, in its information management. We need to take best practices and ideas from other industries and sectors to give us a step change and rapidly evolve our own discipline of quality management into the digital age, otherwise we face extinction.

Artificial intelligence is one pillar of how we can improve and learn, given the volumes of data and information coming at us. AI is just smart software and we need to shape and design it to meet our quality requirements in construction. With the volume of data accumulating project by project, so the development of collaborative Data Trusts from across the construction industry, as well as the individual business AI, will rapidly learn and help to improve performance.

These times can be ferocious in the amount of change and the risk is that new players enter the industry and turn it upside down through much better productivity and higher profits. By using Digital Quality Management at the heart of construction, we can still change the way we design and build to harness the change and ensure that current contractors have a sustainable future and the quality profession has a more influential role in decision- making.

Notes

1 Witzel, M. and Warner, M. (eds), *The Oxford Handbook of Management Theorists* (Oxford: Oxford University Press, 2013).
2 Crosby, P., *Quality Is Free: The Art of Making Quality Certain* (New York. Mentor, 1980).
3 Imai, M., *Kaizen: The Key to Japan's Competitive Success* (New York. McGraw-Hill, 1986).
4 Argyris, C., *Management and Organizational Development: The Path from XA to YB* (New York: McGraw-Hill, 1971).
5 Conan Doyle, A., 'A scandal in Bohemia'. *The Strand Magazine*. (1891). Retrieved from www.gutenberg.org/files/1661/1661-h/1661-h.htm
6 Trudeau, C.R., *The Public Speaks: An Empirical Study of Legal Communication*. (2012). Retrieved from https://works.bepress.com/christopher_trudeau/1/
7 Confucius, *The Analects Book 2*. Trans. J. Legge. (475–221 BCE). Retrieved from https://ebooks.adelaide.edu.au/c/confucius/c748a/index.html

Appendix 1 Quality professional role profile, 2030

A Management systems

1   Leading certification to ISO 9001:2025 that embeds latest technological attributes into processes to measurably create business value.
2   Managing virtual integrated management systems, using Augmented Reality (AR) that provides oral work instructions to guide operatives.
3   Use Machine Learning (ML) to review automated changes to processes for approval.
4   Programme and code process flows linked to tagged knowledge management videos, images, reports and standards for a 'one-stop-shop' of information and knowledge.
5   Programme BIM digital models with a quality management layer for flagging defects and errors, highlighted by as-built laser scans contrasted with the design.

B Auditing, inspections and surveillance

1   Use Virtual Reality (VR) headsets and immersion pods to audit proposed design changes.
2   Use Augmented Reality (AR) to audit as-built on site vs BIM design.
3   Remote Quality Control Inspections at offices and factories of suppliers and contractors using AR headsets to assess on-site issues and review non-conformances.
4   Licensed to fly drones to carry out quality control inspections on site.
5   Material inspections and tests provided by embedded sensors streaming live performance data with automated alerts on non-conformances.

C Information management and reporting

1   Administering Business Intelligence (BI) tools to connect, benchmark and slice & dice data for management reports on KPIs vs actual progress towards project and business objectives.

2   Understand telematics for machine reporting from plant and equipment connected to the BI dashboard.
3   Carry out research and reporting using Artificial Intelligence.

D Teams

1   Use voice-activated AI to do the following:
   - Manage quality team member bespoke training packages.
   - Review performance capabilities.
   - Manage objective setting and reviews.
   - Authorise absences, expenses, recruitment and IT requests.

E Security and business continuity

1   Understand cyber and information threats to continue working while systems are being cleansed.
2   Update management systems in real time for emergency incidents using mobile mapping solutions by on-site operatives.
3   Understand blockchain technology for authorising payments to suppliers and contractors upon completion of work meeting above standards and specification.

F Learning and quality tools

1   Train staff in quality tools (Six Sigma, etc.) remotely in real time on the job through smart glasses/phones/tablets.
2   Customise e-learning packages for all team members following baseline assessment of their quality management knowledge and understanding.
3   Facilitate workshops and Quality Tool Box Talks remotely and understand how to interact with operatives when not in the same physical location.
4   Design quality management training using gamification techniques.
5   Understand crowdsourcing for problem-solving quality issues.

Appendix 2 2030 construction site digital quality

It is another typical day in 2030 for the digital quality professional, Jennifer, in Chester, UK, as she walks towards the security access at the site office and by simply carrying her mobile in her handbag, the automated barriers slide open. From the moment she arrives, cameras follow her every step of the way and ensure she remains safe, using a system introduced just a year before. 'Active intervention' could predict the risk of injury and instruct hovering, safety drones to intervene and protect humans from themselves, other humans and machinery. From 200 construction sites using the system, there had been no deaths or serious injuries and it was hailed as the system to end the horrendous, industrial suffering of avoidable accidents that had gone on for thousands of years.

Over the morning's coffee, Jennifer reviews the quality performance reports automatically sent by the digital business brain for access by her ultra thin, foldout smart device. The reports are generated from sensors throughout the design and construction processes. Quality control data for pipework components being built in a factory in China indicated that defect levels had exceeded prescribed the tolerance limits. The report confirmed that the batch has been re-routed by robots and the software error detected and removed. However, the factory items were firmly on her supplier watch list on the quality performance dashboard.

Her PPE includes slipping on the smart glasses that allow access to data and information through eye gazing. The BIM models, digital British and international standards, internal reports, quality control test scores, design specifications and any other 'document' that was available on the laptop and smartphone of yesteryear, is accessible in real time on her glasses. But it is no longer held in the form of documents but rather appears as digital data. Documented reports are compiled from digital stored pieces located from different places in a database of construction knowledge. Using her glasses she gazes at options that change the colour of her own smart jacket and trousers to the mandatory fluorescent yellow and stiffens the fabric to protect her. She pulls on her safety boots, loaded with sensors to monitor when they need to be replaced and airbags to protect her feet and legs. The kinetic energy, as she walks, charges her mobile wirelessly, so that these days power chargers and cables are no longer in use.

The hard hat has a smorgasbord of sensors such as temperature, carbon monoxide, falls and sensors for identifying falling objects that the safety drones would intervene and catch before they reach her.

Out on site, robots were pouring advanced self-healing concrete and laying artificial bricks in an efficient process, calculated by the construction AI brain. The site had a modest handful of women and men supervising the robots. Electric earth-moving trucks and excavators quietly rumble past Jennifer autonomously, without drivers. The next delivery of components constructed off-site are landing, ready for robots to collect and assemble them. The round building is rapidly forming in front of Jennifer's eyes as she approves the automated witness hold points that have been pre-approved by the AI brain. Embedded sensors are testing the concrete, as it is being poured and matched with the sensor readings from the batch plant, to ensure consistency.

Jennifer was glad to be back on site after the automated alert to her manager a week before, warning that her stress and anxiety levels, which were monitored by her smart watch, had reached unacceptable levels. Her manager had spoken to her and talked through the need to take three days rest. She had reluctantly agreed and her work technology had been disconnected to prevent any temptation. Mental well-being was a priority concern to her employer.

The community centre for the town had been automatically approved from crowdsourcing, across the local population. The Artificial Community Council had noted the uptick in social media discussions by the local residents on needing a meeting place and had then automatically put forward three design options, for further discussion and debate. The vote had come in last month and been approved by a large majority. The design called for most components to be manufactured off-site, and then production commenced immediately in China. The identified site of a former supermarket had been demolished and cleared the day before the construction resources were deployed with autonomous vehicles and drones preparing the ground, ready for the human supervisors to arrive. Jennifer mused that most of the dinosaur supermarkets had disappeared in the same way that Blockbusters video stores, Toys R'Us and BHS had gone shortly before her two mums had celebrated their twentieth wedding anniversary.

The project quality planning had been developed by the AI brain and built into the project management system to inspect and test at every level from the macro of walls down to the nano level of the materials. An alarm sounded in Jennifer's ear and she shifted her gaze on the head up display. A sensor had been triggered for a stuck floor beam that a robot couldn't unjam. It must be more than just a process problem, otherwise the drones would have assisted or a human site supervisor would have been contacted, instead of her. This must be a related quality issue, she thought.

Jennifer spoke quietly into the mobile device and a drone landed for her to step into. The safety vest clipped around her, both enveloping a set of concealed air bags around her and securing her into the seat, before carrying her to the first floor building (humans were no longer allowed to walk around a construction site). She stepped out. The robot forlornly pulling on the beam backed away, patiently waiting for instruction and Jennifer could see the reason for the quality alarm. The end of the beam had been slightly damaged and the red triangle hologram floating above it, showed it was beyond recommended limits but within the contract specification limits. It required a decision on whether to allow it to be placed or not. She visually inspected it but there was only slight scratching and there didn't look like any significant damage. Jennifer's head up display gave her data readings from inside the beam and a probability rating of risk impact to the building design, if she agreed to a concession. The risk was within tolerance based on live global performance data of similar minor damage incidents, over the past ten years and suggested that since it was rated at 97.8 per cent, she could safely agree to it being used. She gazed at the approval button, which turned green, before returning to the drone. Robots rolled into view with a drone to move the beam into position.

As she flew back over the site, she spoke into her mobile to summarise her report while the AI brain would edit video footage, made by the numerous cameras on the machines and those dotted around the site. The report would be created with Jennifer's one-minute oral summary, supporting data and the short video and sent to her manager with a copy automatically filed into the BIM model.

Back in the pop-up site cabin, the ban on paper was making the site manager, Sephora, grumpy. In her late sixties, she had sneaked in a paper notebook and a pen for note taking, which caused a certain level of amusement, before it was spotted on camera by the project manager, Jordan, based in Sydney. Making oral notes on issues that were converted into digital notes for automatic actioning by the AI brain made her feel like she wasn't a proper manager compared to 2020s.

An audit report beeped into her ear bud and Jennifer unfolded her mobile to the size of an old-fashioned tablet. Oral testimony had been gathered from the site team and supply chain, together with photographic, video and laser-scanned evidence. The report was an amber plus and Jennifer frowned. It had been several months since such a poor rating had been found. The root causes were reported as a quality management trainee failing a quality control capability test twice, although he had been predicted to pass first time with flying colours. Jennifer was mentoring Li who lived in China and he was working remotely with her, on the community centre project. She video-called him but he didn't pick up. She spoke into her mobile to create an alert to ping to him to call her and with the AI sensing Jennifer's heightened stress level, which would have rubbed off on Li, it added a smiley face.

Other root causes had been four clustered, first-time inspection failures on specially cast aluminium alloy metals, made in factories in Russia. This had been rectified before the orders had reached the UK but the ripple of breached quality control acceptance levels in the preceding week was unusual, given the excellent track record of the supplier. Jennifer spoke into her mobile asking for suggestions for the failures. The AI brain spoke into her ear; based on probability, it was a sensor misreading the zinc quality. Other options had such low probability that the AI brain provided only one suggestion. Jennifer spoke into her mobile once more, asking for a check on the sensors as part of the pyrometallurgical process investigation. It would be sent to the supplier's quality control manager.

The soft beep indicated a video call and she accepted when she saw it was Li. After the pleasantries, Li volunteered that his young son was keeping him and his wife awake at night, affecting both their concentration levels at work. Li had guessed his test results would affect the audit report and knew there was little point in waiting for Jennifer to raise it first in the conversation. She reassured him that she wanted to help and asked him what he needed to ensure he would pass the test next time, since his quality management duties were suspended until he passed. He suggested a digital work voucher for a robot baby sitter to intervene at night. Jennifer asked Li to hold and she flashed up her quality management budget on her head up display with an instant quote. The project could afford it. She gave Li a thumbs up and said that the robot would attend for two weeks when they would speak again and gave her best wishes to him and his wife.

Over lunch of sandwiches printed by their 4D printer, the small site team gathered around, both face to face and by video, exchanging gossip and updates on the world around them. Aland, their augmented reality architect, who was of Syrian descent, was speaking Arabic over the video from Paris, which was instantly being translated over speakers, about the virtual rebuilding of important landmarks in his homeland, following the end of the war in 2019. Sephora said she had received good news from her nano medic that the handful of cancer cells had been removed and she was in remission, sparking a round of applause. Adaego, videoing in from Nairobi, who was their carbon monitor, told them that her nine-year-old son had asked the previous day if she had ever seen a rhino in the wild and she had described how she had been one of the last people to see one, before they had become extinct roaming free in Africa. Sephora chimed in that there was talk that they would be reintroduced from a DNA zoo breeding programme but Jennifer said it should never have come to this in the first place. The door slid open and David, the project's AI and digital designer, glided in on his hover chair that carried him everywhere. He apologised for being late but had been held up by a Luddite ground driver who had hit a driverless taxi, causing a queue. Luckily no one was hurt, although everyone involved had to be checked by the robot medics despatched to the incident. The team felt David's sensitivity, as they recalled his ground car accident years before leaving him paralysed. He was on the waiting list for a nerve transplant to restore movement to his legs.

Suddenly a klaxon sounded across the site. David retraced his hover path to site, to check on the return of the robot laser scanner. Jennifer looked at Sephora and they both frowned. The video team members hurriedly said their goodbyes and disconnected. All their mobile screens were flashing red; there was an attempted security breach. Outside all machinery had ceased, after the drones had landed in their parking spaces and the autonomous vehicles had crawled back to form an orderly line. The automatic fail safe was programmed in the event of a cyber security compromise. The two women left alone patiently waited and, after ten minutes, the all-clear klaxon sounded with the machines once again moving in all directions. Risk management was at the heart of all actions by the AI brain.

Another beep into her ear and Jennifer could see a request to approve the second floor building completion. She didn't need to visually inspect but she liked to feel in touch with the site, even though there were hardly any humans around. She felt a kindred spirit with the machines that worked tirelessly and had camera eyes that seemed almost sentient in how they looked at her and responded to commands and requests.

Minutes later, standing on the second floor she walked around the rapid construction of pre-cast concrete flooring bolted together and, with all dials on the HUD showing 100 per cent, she approved the second floor construction complete. Blockchain technology instantly paid the subcontractor business (after automatically calculating and deducting government taxes), who was carrying out the work and the fitting out started to commence.

The continuous beeping indicated that the working day was ended for all and Jennifer took a swift look around. Another two days and the community centre would be finished, ready for handover to the residents. She felt a great sense of pride in her digital quality work, as she exited the barriers to start the journey home. She would take a driverless taxi home paid in cryptocurrency, on a dedicated highway, which had no traffic lights and get her home quicker. The thought of traffic jams and near misses with a manually driven car made her shudder.

Appendix 3 Reinforced concrete wall

Quality information example

Use the Quality Information Model to assess the information required for each aspect of People, Processes, Machines and Materials throughout the asset life-cycle to demonstrate the performance quality has been achieved.

For construction of an in situ, reinforced concrete wall, what information might be required?

Client testimonials and references should be used to pull out evidence for each element but should be verified by speaking to the client directly. Table A3.1 sets out examples (this is NOT a definitive list) of the information to be checked, copied and securely stored by the contractor as part of an Approved Supplier List.

Much of this evidence will be replicated across similar activities and processes and hence the information required should be set out across phases of the project and requested either as a follow-up to the bid process (to aid clarification of the bid) or prior to work commencing.

The evidence should ideally be in digital format to be added to an approved supplier list file, stored within a database. The evidence could be sent directly to the contractor by email or, better still, uploaded into a secure cloud that provides confidentiality. Quality audits could also take copies of the evidence.

Table A3.1 Reinforced concrete wall data

Q.I.	Prompt	Information examples
People	Are those who are constructing the wall capable of achieving the specified standard of concrete?	Is there evidence of the subcontractor's operatives meeting the capability? • NVQ Diploma in Formwork (Construction) Level 2 • ICT Certificate in Concrete Technology & Construction: Stages 1–4 • Certificate in Ready Mixed Concrete Technology – Level 4 • NVQ Diploma in Steelfixing Occupations (Construction) Level 2 • CPCS, CISRS, ALLMI and CSCS cards • Competency Assessments demonstrating years of experience as concrete worker • ISO 44001 certification* on collaboration
Processes	Are processes up to date and documented for constructing the wall?	• ISO 9001 certification for QMS • Subcontractor bid document • Risk Assessments and Method Statements, ostensibly for H&S, but also provide evidence of approach and experience, in quality of the work • Procedures and Work Instructions
Machines	Construction plant for lifting and pumping, vibration equipment, total station/laser scanner for setting out, embedded concrete sensors, strength and durability testing	Appropriate calibration certificates and checklists for: • concrete pump • concrete sensors • poker and formwork vibrators • thermo-hygrometer • concrete moisture meters • Windsor probe • ultrasonic pulse velocity • pullout test • rebound hammer
Materials	Concrete mix, steel reinforcement, release agent, construction and contraction joint materials, joint filler and sealant	• Concrete mix design specification • BS EN 206 and BS 8500 concrete certification • CE marking to BS EN 1090 for steel • ISO/IEC 17025 certification (competence of testing and calibration laboratories)

Appendix 4 Summary of the digital learning points

Table A4.1 Digital learning points

Chapter	Subject	Digital learning points
6	Quality Information Model	1. Quality Information Model – identify information on people, processes, machines and materials that leads to required performance 2. Build quality management data and information requirements into the business Master Data Management (MDM) strategy 3. Create a Quality Knowledge Management system using CoPs, CoEs, KM software and Intranet webpages 4. Develop strong Quality Knowledge skills in training, coaching, mentoring, presentations, site TBTs and wider communications
7	Data and Information Management	1. Identify quality attributes specific to project phases and components and then quality management information needed to meet performance requirements 2. Data quality – create and monitor performance measures for data based on accuracy, consistency, completeness, integrity and timeliness 3. Information quality – understand how BIM models are developed and create a Quality Management layer in the models. Create a demand on IT for Quality Management information requirements across the business and in specific projects. 4. Convert traditional Quality Management documents into digital information
8	Business Intelligence and data trusts	1. Influence business objectives and develop quality objectives, KPIs and metrics from them to add value to the business 2. Learn and understand the BI software solutions – Tableau, Microsoft BI, Qlik, Salesforce and Birst 3. Use data visualisation to showcase quality measurement. Explain to business intelligence/IT teams the quality requirements. 4. Ensure BI processes are mapped out as part of the IMS 5. Experiment with data sources for quality – sensors, plant telemetry and RFI tags on materials 6. Create and collaborate with data trusts

Chapter	Subject	Digital learning points
9	Quality Management culture and governance	1. Quality management governance philosophy has three ingredients: decision-making structures, processes and collaboration. 2. Develop a quality culture and benchmark using 'river' survey maps. 3. Quality Management governance structure includes the quality management infrastructure, people talent and organisational culture, oversight and authority and strategic governance. 4. Leverage technology into live reporting and drive metrics to decision-makers that add business value. 5. Use microburst videos, apps and VR for innovative quality management learning customised to different audiences.
10	Digital capabilities	1. Digital capabilities are essential so improve yours and help others to improve ICT proficiency, Information & Data literacy, Digital creation, Digital Communications, Digital learning and Digital well-being. 2. Play with the latest technology – don't be afraid of making mistakes! 3. Encourage the use of video conferencing, calls and undertake online presentations to reach a wider audience. Record them and make them available later for those who cannot join in. 4. Become a remote online expert with surgeries at set times to give everyone an opportunity to reach you with their problems and issues. 5. Use short sharp campaigns across multiple platforms to push Quality Management messages.
11	Web-based process management	1. Develop a Digital Construction Management System (DCMS) to create a more dynamic approach to integrate knowledge, workflows, HUD instructions and BIM models around robust process architecture. 2. Develop a User Interface for the management system, which is user-friendly, intuitive and easy to navigate. 3. Use SIPOC methodology to develop process maps starting with top-level processes and working down to the lowest processes. Ensure that all processes connect with outputs flowing into inputs to build a true IMS. 4. Embed hyperlinks in the IMS to connect to Quality Knowledge Management systems. 5. Use workflow software for processes where it makes life easier for the users and pares down procedural text. 6. Investigate HUD technology for work instructions in higher-risk tasks on-site. 7. Connect BIM models with layers of active Quality/ Environment/H&S/Security information. 8. Monitor usage to find heat maps and dead areas of the IMS. Use analytics to drive improvements. 9. Connect IMS into other stakeholder management systems.

Chapter	Subject	Digital learning points
12	Drones	1. Qualify as a drone pilot. 2. Be clear: what is the image data to be used for? Is there a clear process to quality assure the data? 3. Use drones for quality control inspections on site. Package up results with narrative report and drone data, including laser scans and video and photographic images. 4. Assess and track the transfer points of the drone data to assure quality. Does the manufacturer's user manual provide recommendations on data management? 5. On audits, check: the drone user manual is version-controlled and maintained, the drone batteries are at the optimum, and drone servicing is carried out 6. Assess data security for any thumb drives or wi-fi used to download data. 7. Do drone dashboards show and record telemetries? Can this data provide improvements for future use?
13	Construction plant: autonomous vehicles and telemetry	1. Assessing the capability of a driver for Level 2 autonomy, who may be based on the other side of the world in an office from where the vehicle is operating, will require demonstrable evidence other than a driving licence. Will a new 'gaming'-type driving qualification be required? But if the vehicle is a Level 3 or 4 and self-driving, then the capability of the AI becomes more appropriate than a human capability. 2. Understand what telemetry data is available for each type of construction plant. Does any of the data report useful quality performance? If it does, then check that it is being monitored by the plant contractor and utilised. 3. When autonomous machines are self-driving, only the AI process in its very basic sense could be mapped. How will processes be assessed with an AI 'black box' of tasks between an input and output? 4. Does the telemetry have documented processes for data quality? How can the operator be sure that the telemetry measurements and information are accurate? 5. Are the machines being properly maintained? Many of the checks will focus on safety but some will report on serviceability. Downtime is waste. How are the plant contractors trained on the plant before operations? 6. When materials are being moved, is the plant protecting those materials? For example, lifting bricks or blocks by crane or hoist. Are construction materials being moved efficiently by the plant?

Chapter	Subject	Digital learning points
14	Robotics, lasers and 3D printing	1. Understand how robots, laser scanners and 3D printers are operated. Does any of the performance data report on quality performance? If it does, then check that it is being monitored by the robot or 3D printer contractor and used to improve performance. 2. Does the use of the robots, lasers and 3D printers have documented processes? Are user manuals available to be referenced to documented procedures? Are there links to current user manuals? How can the operator be sure that the robots and 3D printers will deliver to the required quality performance? 3. How is registration of laser scans quality managed? Is there an audit trail of changes to the digital point cloud models through saved scans? 4. Are the robots, laser scanners and 3D printers being properly maintained? Many of the checks will focus on safety but some will report on serviceability. How are the robots, laser scanners and 3D printers being checked before operations? 5. When materials are being pumped, is the robot suitably protecting those materials? 6. Has the data for laser scanners, robots and 3D printers been quality assured? Are the scans that are sent to stakeholders fit for purpose to meet agreed specifications? Is the scanned data secured and available as part of the digital asset management? 7. Can robots, laser scanners and 3D printers demonstrate they are within specified calibration requirements? Are there calibration records available prior to the scan being created? 8. Are robot and 3D printer supervisors and laser scanner operators suitably competent? Can training/qualifications be demonstrated?
15	AR, MR and VR	1. Augmented Reality (AR) has a great capacity to add functionality and insight into the as-built world vs the designed BIM model, identifying errors and defects and providing Cost of Quality to be calculated. 2. Combining technologies such as AR and smart glasses allow eye gaze to call up work instructions on site and in the factory. Check that such QMS 'documents' maintain version control. 3. Virtual Reality (VR) offers simulated Quality Management training such as identifying snagging. Identify scenarios for added value training and find cost-effective suppliers. Lobby to develop such training. 4. Promote VR for clients to view designs and construction scenarios and measure customer satisfaction, before and after

Chapter	Subject	Digital learning points
16	Wearable and voice-controlled technology	1. Experiment with wearable and voice-controlled technology to improve efficiency and effectiveness in the workplace. Consider using IBM Watson Assistant or Alexa for Business. 2. Use eye tracking eye wear, such as Tobii Pro, to research and develop training in Quality Management. Which defects or errors have been missed? 3. Investigate uses for smart glasses by DAQRI to test out accessing management system 'documents' and tagging defects and queries 4. Think about comfort and ease of use of this technology, especially over a long day. Ensure processes for optimising the use are thought out and documented. Eye strain from smart glasses may become the new 'repetitive injury' in coming years. 5. What can the quality profession learn from manufacturing on the application of wearables? Smart glasses, watches, gloves and exoskeletons have industrial uses and we need to assess and develop them to exploit quality management capability.
17	Blockchains	1. Blockchains are a digital ledger of transactions and single source of truth. 2. Using blockchain technology, inspections could trigger payments to a supplier or subcontractor. Quality professionals need to appreciate the impact and potential pressure placed on them to accept or reject the work, given the new financial impact. 3. Blockchains can be used to reliably verify the origins of materials through the supply chain. 4. Used with sensors, blockchains can track how materials and construction components are handled through transportation and storage. 5. Blockchains can provide a new robust approach to auditing with digital signatures providing greater transparency and accountability.
18	Artificial intelligence	1. Appreciate the differences between different types of AI: Artificial Narrow Intelligence (ANI) = Weak AI, Artificial General Intelligence (AGI) = Strong AI and Artificial Super Intelligence (ASI). 2. AI used in construction is connecting data and machine learning to predict project risks (e.g. Autodesk Project IQ). 3. Risk management is aided by smartvid.io-type software that geo-tags features and then learns to identify Quality Management trends and patterns for defects. 4. The next edition of the ISO standard, ISO 9001:2025, will need to cross the AI bridge, so start to think about the impact on future certification. 5. AI's prediction and forecasting ability will rise rapidly over the coming years.

Chapter	Subject	Digital learning points
19	Advanced materials science	1. Wireless sensors and software can provide an alternative to traditional quality control testing of concrete and brickwork. Get to know how these data streams meet project specifications. 2. New advanced materials such as Graphene concrete, new grades of steel, nanotechnology, aerogels and carbon nanotubes will appear in specifications and quality professionals need to understand how such specifications are been met. Are manufacturers' instructions being followed in transportation, handling, storage and installation? 3. Technologies such as photovoltaic road surfaces and smart flooring and roadways, offer new functionality, and quality control testing at installation may be unorthodox or more sophisticated than traditional concrete and asphalt surfaces. 4. Smart cities will emerge with combinations of technologies that may have multiple impacts on quality controls. Using two technologies or advanced materials together may affect stated performance compared to using them separately, requiring a risk assessment.

References

3Dnatives (2018). 'The 3D printing construction market is booming'. 26 January. Retrieved from www.3dnatives.com/en/3d-printing-construction-240120184/

3D Repo (2017). 'Dynamic virtual reality for customer engagement and staff training in construction'. November. Retrieved from http://3drepo.org/wp-content/uploads/2017/11/Dynamic-VR_a4booklet.pdf

Adams, M. (2018). 'Top 10 Machu Picchu secrets'. *National Geographic*. November. Retrieved from www.nationalgeographic.com/travel/top-10/peru/machu-picchu/secrets/ (accessed 31 July 2018).

Akponeware, A.O. and Adamu, Z.A. (2017). 'Clash detection or clash avoidance? An investigation into coordination problems in 3D BIM'. 21 August 2017. Retrieved from www.mdpi.com/2075-5309/7/3/75/pdf

Allen, W.C. 'History of slave laborers in the construction of the US Capitol'. Retrieved from https://emancipation.dc.gov/sites/default/files/dc/sites/emancipation/publication/attachments/History_of_Slave_Laborers_in_the_Construction_of_the_US_Capitol.pdf (accessed 1 June 2005).

Allied Market Research 'Construction lasers market by product'. Retrieved from www.alliedmarketresearch.com/construction-lasers-market (accessed September 2018).

All Party Parliamentary Group for Excellence in the Built Environment (APPGEBE) (2016). 'More homes, fewer complaints'. London. Retrieved from https://policy.ciob.org/wp-content/uploads/2016/07/APPG-Final-Report-More-Homes-fewer-complaints.pdf

Angus, W. and Stocking, L.S. (2017). 'VR transforms doctors, nurses, and staff into virtual construction allies'. Autodesk. 1 November. Retrieved from www.autodesk.com/redshift/vr-construction/

ArchDaily (2017). 'Scientists uncover the chemical secret behind Roman self-healing underwater concrete'. 5 July. Retrieved from www.archdaily.com/875212/scientists-uncover-the-chemical-secret-behind-roman-self-healing-underwater-concrete

Argyris, C. (1971). *Management and Organizational Development: The Path from XA to YB*. New York: McGraw-Hill.

ariot.io 'Finally, Augmented Reality for the building lifecycle'. Retrieved from www.ariot.io

ASQ Kaushik, S.K.V. (2017). 'Virtual reality for quality'. June. Retrieved from http://asq.org/2017/06/lean/virtual-reality-vr-for-quality.pdf

Autodesk (2016). 'BIM 360 Project IQ'. Retrieved from https://knowledge.autodesk.com/support/bim-360/learn-explore/caas/video/youtube/watch-v-nuWlpqevfIk.html

AWS 'Alexa for Business'. Retrieved from https://aws.amazon.com/alexaforbusiness/

Balfour Beatty (2016). 'Building the future with 3D printing'. 16 November. Retrieved from www.youtube.com/watch?time_continue=86&v=EogNa8LAWQg

Baron de Bode, C.A. (1845). *Travels in Luristan and Arabistan*, vol. 1, p. 171. Retrieved from https://books.google.co.uk/books?id=i_gqUpmQRIwC&pg=PA97&source=gbs_toc_r&cad=4#v=onepage&q&f=false

Bartlett, C. (2014). *The Design of the Great Pyramid of Khufu*. Retrieved from https://link.springer.com/content/pdf/10.1007%2Fs00004-014-0193-9.pdf. (accessed 14 May 2014).

BBC News (2017). 'Hammond: Driverless cars will be on UK roads by 2021'. 17 November. Retrieved from www.bbc.co.uk/news/business-42040856

BBC News (2018). 'The world's first family to live in a 3D-printed home'. 6 July. Retrieved from www.bbc.co.uk/news/technology-44709534

BDK Daizokyo Text Database (n.d.). *Pāli Tripitaka*. B2025, Ch6 P135, *The Baizhang Zen Monastic Regulations*. Trans. Shohei Ichimura. Retrieved from http://21dzk.l.u-tokyo.ac.jp/BDK/bdk_search.php?key=construction&strct=1&kwcs=50&lim=50.

Benning, W. (1804; trans. 1827). *Code Napoleon*, Retrieved from http://files.libertyfund.org/files/2353/CivilCode_1566_Bk.pdf

Bentley, M.J.C. (1981). *Quality Control on Sites*. Watford: BRE.

Blanco, J.L., Fuchs, S., Parsons, M. and Ribeirinho, M.J. (2018) 'Artificial intelligence: construction technology's next frontier'. McKinsey. Retrieved from www.mckinsey.com/industries/capital-projects-and-infrastructure/our-insights/artificial-intelligence-construction-technologys-next-frontier

Blueprint robotics 'A better way to build'. Retrieved from www.blueprint-robotics.com

Booth, B. (2018). 'Slightly heavier than a toothpick, the first wireless insect-size robot takes flight', CNBC News, 3 November. Retrieved from www.cnbc.com/2018/11/02/about-the-weight-of-a-toothpick-first-wireless-robo-insect-takes-off.html

BRE Group (2018). 'Blockchain feasibility and opportunity assessment 2018'. January. Retrieved from https://bregroup.com/wp-content/uploads/2018/02/99330-BRE-Briefing-Paper-blockchain-A4-20pp-WEB.pdf

Brickschain 'Connect everything to everyone'. Retrieved from www.brickschain.com

British Research Station (1960). *National Building Studies Special Report 33: A Qualitative Study of Some Buildings in the London Area*. Watford: BRE.

Brokk Inc. 'The smart power lineup'. Retrieved from www.brokk.com/us/

Brunel Museum. (2018). 'The Thames Tunnel'. Retrieved from www.brunel-museum.org.uk/history/the-thames-tunnel

BSI (2012). *BS ISO 10018:2012 Quality Management: Guidelines on People Involvement and Competence*. Milton Keynes: BSI Standards Limited.

BSI (2015a). *BS EN ISO 9001:2015 Quality Management Systems: Requirements*. Milton Keynes: BSI Standards Limited, Section 1, p. 1.

BSI (2015b). *BS EN ISO 9000:2015 Quality Management Systems: Fundamentals and Vocabulary*. Milton Keynes: BSI Standards Limited.

BSI (2015c). *BS EN ISO 9001:2015 Quality Management Systems: Requirements*. Milton Keynes: BSI Standards Limited, Section 1, p. 1.

BSI (2017). 'BS 11000 has been replaced by ISO 44001 Collaborative Business Relationships Management System'. Retrieved from www.bsigroup.com/en-GB/iso-44001-collaborative-business-relationships/

BSI (2018a). *BS ISO 31000:2018 Risk Management: Guidelines*. Milton Keynes: BSI Standards Limited.

BSI (2018b). *BS EN ISO 10005–2018 Quality Management: Guidelines for Quality Plans*. Milton Keynes: BSI Standards Limited, Section 3.2, p. 2.

BSI *Training Courses for ISO 9001 Quality Management*. Retrieved from www.bsigroup. com/en-GB/iso-9001-quality-management/iso-9001-training-courses/?creative= 194426026494&keyword=iso%209001%20course&matchtype=p&network=g&devic e=c&gclid=Cj0KCQjwgOzdBRDlARIsAJ6_HNnegv8lMukZ2lDkUzIAtug-hpa07zbY6-ajRuv53lDJGh2eyUBTWYsaAmJrEALw_wcB

Building in Quality Working Group (2018). *Building in Quality: A Guide to Achieving Quality and Transparency in Design and Construction*. Retrieved from www.architecture. com/-/media/files/client-services/building-in-quality/riba-building-in-quality-guide-to-using-quality-tracker.pdf

Building System Planning Inc. 'ClashMEP'. Retrieved from https://buildingsp.com/index. php/products/clashmep

Business Insider (2014). 'A venture capital firm just named an algorithm to its board of directors – here's what it actually does'. 23 May. Retrieved from www.businessinsider. com/vital-named-to-board-2014-5?IR=T

CAA (2015). *Permissions and Exemptions for Commercial Work Involving Small Drones*. Retrieved from www.caa.co.uk/Commercial-industry/Aircraft/Unmanned-aircraft/Small-drones/Permissions-and-exemptions-for-commercial-work-involving-small-drones

Calkins, R.G. (1998). *Medieval Architecture in Western Europe: From* A.D. *300 to 1500*. New York: Oxford University Press.

Chandler D. (2011). 'Self-healing material can build itself from carbon in the air'. Cambridge, MA: MIT. 11 October. Retrieved from http://news.mit.edu/2018/self-healing-material-carbon-air-1011

Changali, S., Mohammad, A., and van Nieuwland, M. (eds) (2015). 'The construction productivity imperative;' McKinsey Global Institute. Retrieved from www.mckinsey. com/industries/capital-projects-and-infrastructure/our-insights/the-construction-productivity-imperative (accessed 14 July 2015).

CIOB BIM+ (2017). 'Komatsu takes first step to the autonomous construction site'. 17 December. Retrieved from www.bimplus.co.uk/news/komatsu-takes-first-step-autonomous-construction-s/

Cisco & Oxford Economics (2017). 'Modeling to inform the future of work'. Retrieved from www.cisco.com/c/dam/assets/csr/pdf/modeling-inform-future-work.pdf

CISION PR Newswire (2018). 'Global Augmented Reality (AR) and Virtual Reality (VR) market is forecast to reach $94.4 billion by 2023 – soaring demand for AR & VR in the retail & e-commerce sectors'. 31 July. Retrieved from www.prnewswire.com/ news-releases/global-augmented-reality-ar-and-virtual-reality-vr-market-is-forecast-to-reach-94-4-billion-by-2023-soaring-demand-for-ar-vr-in-the-retail-e-commerce-sectors-300689154.html

Cognizant (2017). '21 jobs of the future: a guide to getting and staying employed over the next 10 years'. White Paper. Retrieved from www.cognizant.com/whitepapers/21-jobs-of-the-future-a-guide-to-getting-and-staying-employed-over-the-next-10-years-codex3049.pdf.

Colas Group, WattWay (2016). 'Wattway, the Colas Solar Road'. 10 November. Retrieved from www.youtube.com/watch?time_continue=2&v=OI9fSnBig3s

Collins English Dictionary 'Data'. Retrieved from www.collinsdictionary.com/dictionary/ english/governance

Conan Doyle, A. (1891). 'A scandal in Bohemia'. *The Strand Magazine*. Retrieved from www.gutenberg.org/files/1661/1661-h/1661-h.htm

Confucius (475–221 BCE). *The Analects*, Book 2. Trans. J. Legge. Retrieved from https:// ebooks.adelaide.edu.au/c/confucius/c748a/index.html

Construction Manager (2018). 'Construction bodies call for an end to retentions'. 24 January. Retrieved from www.constructionmanagermagazine.com/news/construction-bodies-call-end-retentions/

Construction robotics 'SAM100'. Retrieved from www.construction-robotics.com/sam100/

Cozzo, G. (1971). *Il Colosseo*. Rome: Palombi.

CQI *The CQI Competency Framework*. Retrieved from www.quality.org/knowledge/cqi-competency-framework

CQI 'CQI Training Certificates in Quality Management'. Retrieved from www.quality.org/CQI-training-certificates-in-Quality-Management.

Crosby, P. (1980). *Quality Is Free: The Art of Making Quality Certain*. New York: Mentor.

Crossrail Learning Legacy (2017). 'Augmented Reality trials at Crossrail'. 14 March. Retrieved from https://learninglegacy.crossrail.co.uk/documents/augmented-reality-trials-crossrail/

Crow, M. and Olson, C.C. (1966). *Chaucer Life-records*. Oxford: Oxford University Press.

Daily Telegraph (2018). 'Drone near-misses triple in two years', 19 March. Retrieved from www.telegraph.co.uk/news/2018/03/19/drone-near-misses-triple-two-years

DAQRI, 'Smart glasses'. Retrieved from https://daqri.com/products/smart-glasses/

Dauth, W., Findeisen, S., Südekum. J., and Woessner, B. (2017). 'German robots: the impact of industrial robots on workers'. Retrieved from ec.europa.eu/social/BlobServlet?docId=18612&langId=en

Davis, W. (2018). 'SIPOC management: you're in charge. Now what?' *Quality Digest*, 20 August. Retrieved from www.qualitydigest.com/inside/management-article/sipoc-management-you-re-charge-now-what-082018.html

Daw, T. (2013). 'How many stones are there at Stonehenge?' Retrieved from www.sarsen.org/2013/03/how-many-stones-are-there-at-stonehenge.html

Department for Business, Innovation & Skills (2013). *Construction 2025: Strategy*. 2 July. Retrieved from www.gov.uk/government/publications/construction-2025-strategy

Dimov, D., Amit, I., Gorrie, O., Barnes, M., Townsend, N., *et al.*, (2018). 'Ultrahigh performance nanoengineered Graphene–concrete composites for multifunctional applications'. Retrieved from https://onlinelibrary.wiley.com/doi/full/10.1002/adfm.201705183

DoT/CAA (2018). 'New drone laws bring added protection for passengers'. 30 May. Retrieved from www.gov.uk/government/news/new-drone-laws-bring-added-protection-for-passengers

Downing, A.J. and Wightwick, G. (1847). *Hints to Persons about Building in the Country*. New York.

Doxel 'Artificial intelligence for construction productivity'. Retrieved from www.doxel.ai

DroneDeploy (2018a) 'Drones raise the bar for roadway pavement inspection'. Blog, 2 August. Retrieved from https://blog.dronedeploy.com/drones-raise-the-bar-for-roadway-pavement-inspection-9c0079465772

DroneDeploy (2018b). *2018 Commercial Drone Industry Trends Report*. Retrieved from www.dronedeploy.com/resources/ebooks/2018-commercial-drone-industry-trends-report/

Dubai Future Foundation (2018). *Office of the Future*. Retrieved from www.officeofthefuture.ae/#

Dutch Government 'Blockchain projects'. Retrieved from www.blockchainpilots.nl/results

Dyble, J. (2018). 'Understanding SAE automated driving: Levels 0 to 5 explained'. *Gigabit*. 23 April. Retrieved from www.gigabitmagazine.com/ai/understanding-sae-automated-driving-levels-0-5-explained

EC&M (2018). '3D visualization brings a new view to job sites'. 1 October. Retrieved from   www.ecmweb.com/neca-show-coverage/3d-visualization-brings-new-view-job-sites

EksoBionics 'Power without pain'. Retrieved from https://eksobionics.com/eksoworks/

eMarketeer (2017). 'Alexa, say what?! Voice-enabled speaker usage to grow nearly 130% this year'. 8 May. Retrieved from www.emarketer.com/Article/Alexa-Say-What-Voice-Enabled-Speaker-Usage-Grow-Nearly-130-This-Year/1015812

Empire State Realty Trust. 'Empire State Building fact sheet'. Retrieved from www.esbnyc.com/sites/default/files/esb_fact_sheet_4_9_14_4.pdf

Emporis. Philadelphia City Hall. Retrieved from www.emporis.com/buildings/117972/philadelphia-city-hall-philadelphia-pa-usa.

Encyclopaedia Britannica 'Parthenon'. Retrieved from www.britannica.com/topic/Parthenon.

Encyclopedia of the Social Sciences (1938). 'Guilds', vol. VII. New York, pp. 204–224. Retrieved from https://archive.org/details/encyclopaediaoft030467mbp/page/n3

Epson 'A bright horizon for FPV'. Retrieved from www.epson.co.uk/products/see-through-mobile-viewer/moverio-bt-300/drone-piloting-accessory

Epson Moverio (2018). 'DJI MAVIC AIR with Epson Moverio BT300 smart glasses – review part 2 – flight test, pros & cons'. 28 March. Retrieved from www.youtube.com/watch?v=7u5NvNOdSeg

ETH Zürich (2018). 'Knitted concrete'. Retrieved from www.youtube.com/watch?v=spPpkPHK7Q0&feature=youtu.be

Everett, A. (2001). *Cicero: A Turbulent Life*. London: John Murray.

FBR 'Robotic construction is here'. Retrieved from www.fbr.com.au/view/hadrian-x

Federal Aviation Administration (FAA), Dronezone. *Welcome to the FAA DroneZone*. Retrieved from https://faadronezone.faa.gov/#/

Fisch, K. (2010). *Shift Happens* video. 12 July. Retrieved from www.youtube.com/watch?time_continue=21&v=SBwT_09boxE

Frontier Economics (2018). 'The-impact-of-AI-on-work', p. 32. Retrieved from https://royalsociety.org/~/media/policy/projects/ai-and-work/frontier-review-the-impact-of-AI-on-work.pdf

Gartner (2015). 'Market trends: voice as a UI on consumer devices — what do users want?' 2 April. Retrieved from www.gartner.com/doc/3021226/market-trends-voice-ui-consumer

Gershgorn, D. (2018). 'AI is the new space race: Here's what the biggest countries are doing'. *Quartz*. 2 May. Retrieved from https://qz.com/1264673/ai-is-the-new-space-race-heres-what-the-biggest-countries-are-doing/

Ghose, T. (2012). 'Mystery of Angkor Wat Temple's huge stones solved'. *Livescience*, 31 October. Retrieved from www.livescience.com/24440-angkor-wat-canals.html

Gil, L. (2017). 'How China has become the world's fastest expanding nuclear power producer'. Vienna: IAEA. Retrieved from www.iaea.org/newscenter/news/how-china-has-become-the-worlds-fastest-expanding-nuclear-power-producer (accessed 25 October 2017).

Gomes, H. (1996). *Quality Quotes*. Milwaukee, WI: ASQC Quality Press.

Grenfell Tower Inquiry. Retrieved from www.grenfelltowerinquiry.org.uk

GuardHat 'Overprotective in a good sort of way'. Retrieved from www.guardhat.com

Guardian (2017). 'Give robots an "ethical black box" to track and explain decisions, say scientists'. 19 July. Retrieved from www.theguardian.com/science/2017/jul/19/give-robots-an-ethical-black-box-to-track-and-explain-decisions-say-scientists

Guo, Q. (2000). *Tile and Brick Making in China: A Study of the Yingzao Fashi*. Retrieved from www.arct.cam.ac.uk/Downloads/chs/final-chs-vol.16/chs-vol.16-pp.3-to-11.pdf

Hall, W. and Pesenti, J. (2017). 'Growing the Artificial Intelligence industry in the UK'. Retrieved from https://assets.publishing.service.gov.uk/government/uploads/system/uploads/attachment_data/file/652097/Growing_the_artificial_intelligence_industry_in_the_UK.pdf

HM Government (2017). 'Industrial strategy-building: a Britain fit for the future'. White Paper. CM9529. Retrieved from https://assets.publishing.service.gov.uk/government/uploads/system/uploads/attachment_data/file/664563/industrial-strategy-white-paper-web-ready-version.pdf

HoloBuilder 'Full HoloBuilder feature overview'. Retrieved from www.holobuilder.com/features/

Honda (2018). 'Honda walking assist obtains medical device approval in the EU'. 18 January. Retrieved from https://world.honda.com/news/2018/p180118eng.html

Honda 'ASIMO innovations'. Retrieved from http://asimo.honda.com/innovations/

Honda 'ASIMO'. Retrieved from http://asimo.honda.com

HSE (2017). *Health and Safety Statistics for the Construction Sector in Great Britain, 2017*. London: Health and Safety Executive.

Huynh, T. and Jones, M. (2018). 'Scientists create new building material out of fungus, rice and glass'. Retrieved from https://phys.org/news/2018-06-scientists-material-fungus-rice-glass.html

IBIS World (2017). 'Title insurance in the US: US industry market research report'. September. Retrieved from www.ibisworld.com/industry-trends/specialized-market-research-reports/advisory-financial-services/specialist-insurance-lines/title-insurance.html

IBM 'The artificial intelligence effect on industrial products'. Retrieved from https://public.dhe.ibm.com/common/ssi/ecm/17/en/17013217usen/industrial-products-ai_17013217USEN.pdf

IBM 'Watson Assistant'. Retrieved from www.ibm.com/watson/ai-assistant/ (accessed 12 April 2018).

ICAEW 'What is corporate governance?' Retrieved from www.icaew.com/technical/corporate-governance/uk-corporate-governance/does-corporate-governance-matter

ICON (2018). 'Welcome to the future of human shelter'. Retrieved from www.iconbuild.com

IDC (2014). *The Digital Universe of Opportunities*. April. Retrieved from www.emc.com/leadership/digital-universe/2014iview/executive-summary.htm

Imai, M. (1986). *Kaizen: The Key to Japan's Competitive Success*. New York. McGraw-Hill.

IMG. 'Liquid granite'. Retrieved from www.img-limited.co.uk/product/liquid-granite/

Inscriptiones Graecae (2013). Trans. S. Lambert, J. Blok, and R. Osborne. Retrieved from www.atticinscriptions.com/inscription/IGI3/35

ISO (2017). *Survey of Certifications to Management System Standards: Full Results*. Geneva: ISO. Retrieved from https://isotc.iso.org/livelink/livelink?func=ll&objId=18808772&objAction=browse&viewType=1

IW (2016). 'Trial uses drone to carry out Crossrail shaft inspections'. 21 November. Retrieved from www.infoworks.laingorourke.com/innovation/2016/october-to-december/trial-uses-drone-to-carry-out-crossrail-shaft-inspections.aspx

Japan Times (2017). 'Candy-carrying drone crashes into crowd, injuring six in Gifu', 6 November. Retrieved from www.japantimes.co.jp/news/2017/11/05/national/candy-carrying-drone-crashes-crowd-injuring-six-gifu/#.W2lpDy2ZOu4

JISC 'Building digital capabilities: The six elements defined digital capability model'. Retrieved from http://repository.jisc.ac.uk/6611/1/JFL0066F_DIGIGAP_MOD_IND_FRAME.PDF

Josephson, P-E. (1998). 'Defects and defect costs in construction: A study of seven building projects in Sweden'. Working Paper, Department of Management of Construction and Facilities, Chalmers University of Technology, Gothenburg, Sweden. Retrieved from http://publications.lib.chalmers.se/records/fulltext/201455/local_201455.pdf

Karczewski. T. (2018). 'IFA 2018: Alexa devices continue expansion into new categories and use cases'. Alexa Blogs. 2 September. Retrieved from https://developer.amazon.com/blogs/alexa/post/85354e2f-2007-41c6-b946-5a73784bc5f3/ifa-2018-alexa-devices-continue-expansion-into-new-categories-and-use-cases

Kim, H., Haas, C.T. and Rauch, A.F. (2003). 'AI based quality control of aggregate production'. Retrieved from www.semanticscholar.org/paper/Artificial-Intelligence-Based-Quality-Control-of-Kim-Haas/6c0d58e953b8546fb9f86223c6ebefd78df6d423

Kinsella, B. (2018). 'Amazon Alexa now has 50,000 skills worldwide, works with 20,000 devices, used by 3,500 brands'. Voicebot.ai. 2 September. Retrieved from https://voicebot.ai/2018/09/02/amazon-alexa-now-has-50000-skills-worldwide-is-on-20000-devices-used-by-3500-brands/

Kloberdanz, K. (2017). 'Smart specs: OK glass, fix this jet engine'. *GE Aviation*. 19 July. Retrieved from www.ge.com/reports/smart-specs-ok-glass-fix-jet-engine/

Komatsu (2017). 'Komatsu intelligent machine control: the future today'. Retrieved from www.komatsu.eu/en/Komatsu-Intelligent-Machine-Control

Koskela, L. (1996). 'Towards the theory of (lean) construction'. Retrieved from https://pdfs.semanticscholar.org/8e87/bc1a102603e9decedf4bb4650803c90f94e4.pdf

kununu 'What is company culture? 25 business leaders share their own definition'. Blog. Retrieved from https://transparency.kununu.com/leaders-answer-what-is-company-culture/ (accessed 31 March 2017).

Laing O'Rourke 'Heathrow Terminal 5, London, UK'. Retrieved from www.laingorourke.com/our-projects/all-projects/heathrow-terminal-5.aspx

Leica and Autodesk (2015). 'When to use laser scanning in building construction'. Retrieved from http://constructrealityxyz.com/test/ebook/LGS_AU_When%20to%20Use%20Laser%20Scanning.pdf

Leviathan, Y. and Matias, Y. (2018). 'Google Duplex: an AI system for accomplishing real-world tasks over the phone'. Google AI Blog. 8 May. Retrieved from https://ai.googleblog.com/2018/05/duplex-ai-system-for-natural-conversation.html

LHR Airports Limited 'Heathrow facts & figures'. Retrieved from www.heathrow.com/company/company-news-and-information/company-information/facts-and-figures

Lillian Goldman Law Library (2008). *Code of Hammurabi*. Trans., L.W. King. Retrieved from http://avalon.law.yale.edu/ancient/hamframe.asp

Liu, J. (2018). 'The difference between AR, VR, MR, XR and how to tell them apart'. Hackernoon, 2 April. Retrieved from https://hackernoon.com/the-difference-between-ar-vr-mr-xr-and-how-to-tell-them-apart-45d76e7fd50

Logan, M. ([1912] 1972). *The Part Taken by Women in American History*. New York: Arno Press.

Loucks, J., Davenport, T. and Schatsky, D. (2018). 'Early adopters combine bullish enthusiasm with strategic investments'. In *State of AI in the Enterprise*, 2nd edn. Deloitte. Retrieved from www2.deloitte.com/insights/us/en/focus/cognitive-technologies/state-of-ai-and-intelligent-automation-in-business-survey.html

Lucideon Limited (2014). 'Brickwork evaluation of Battersea Power Station'. Retrieved from www.lucideon.com/construction/insight-hub/case-studies/brickwork-evaluation-of-battersea-power-station

Lucon, O., Ürge-Vorsatz, D. *et al.* (2014). 'Buildings', in IPCC, *Climate Change 2014: Mitigation of Climate Change*. Cambridge, Cambridge University Press. Retrieved from www.ipcc.ch/pdf/assessment-report/ar5/wg3/ipcc_wg3_ar5_chapter9.pdf

MAA, BAPLA, DoT (2016). *Small Remotely Piloted Aircraft Systems (Drones) Mid-Air Collision Study*. Retrieved from https://assets.publishing.service.gov.uk/government/uploads/system/uploads/attachment_data/file/628092/small-remotely-piloted-aircraft-systems-drones-mid-air-collision-study.pdf

Mace 'Construction productivity: the size of the prize'. Retrieved from www.macegroup.com/perspectives/180125-construction-productivity-the-size-of-the-prize (accessed 24 January 2018).

Magic Leap 'Free your mind'. Retrieved from www.magicleap.com

Magic-plan 'Create a floor plan within seconds'. App. Retrieved from www.magic-plan.com

McDonald, R. (2015). *Root Causes and Consequential Cost of Rework*. Catlin Insurance North America Construction.

McKinsey & Co (2018). 'What AI can and can't do (yet) for your business'. Retrieved from www.mckinsey.com/business-functions/mckinsey-analytics/our-insights/what-ai-can-and-cant-do-yet-for-your-business

McPartland, R. (2016). 'What is the Common Data Environment (CDE)?' *NBS*, 18 October. Retrieved from www.thenbs.com/knowledge/what-is-the-common-data-environment-cde

Medium, 'Introducing artificial intelligence for construction productivity'. Retrieved from https://medium.com/@doxel/introducing-artificial-intelligence-for-construction-productivity-38a74bbd6d07 (accessed 24 January 2018).

Meisner, G. (2013). 'The Parthenon and Phi, the Golden Ratio'. Retrieved from www.goldennumber.net/parthenon-phi-golden-ratio/ (accessed 20 January 2013).

Milgram, P., Takemura, H., Utsumi, A. and Kishino, F. (1994). 'Augmented Reality: A class of displays on the reality-virtuality continuum'. Retrieved from http://etclab.mie.utoronto.ca/publication/1994/Milgram_Takemura_SPIE1994.pdf

MISIS (2017) 'Metalloinvest and NUST MISIS to launch laboratory for development of new steel grades'. Retrieved from http://en.misis.ru/university/news/science/2018-01/5143/

Moore, S., 'Gartner says worldwide business intelligence and analytics market to reach $18.3 billion in 2017'. Retrieved from www.gartner.com/newsroom/id/3612617 (accessed 17 February 2017).

Mosher, D. (2011). 'It's official: Stonehenge stones were moved 160 miles'. *National Geographic Magazine*, 24 December. Retrieved from https://news.nationalgeographic.com/news/2011/12/111222-stonehenge-bluestones-wales-match-glacier-ixer-ancient-science/

MX3D (2018). 'MX3D bridge'. September. Retrieved from https://mx3d.com/projects/bridge/

National History Foundation. '"Our Ural". Kasli cast iron pavilion'. Retrieved from https://nashural.ru/article/istoriya-urala/kaslinskij-chugunnyj-pavilon/ (accessed 23 January 2016).

NBS (2017). 'BIM dimensions – 3D, 4D, 5D, 6D BIM explained'. Retrieved from www.thenbs.com/knowledge/bim-dimensions-3d-4d-5d-6d-bim-explained

NCCR Digital Fabrication (2015). 'In situ fabricator'. 18 June. Retrieved from www.youtube.com/watch?v=loFSmJO3Hhk

NCCR Digital Fabrication (2017). 'In situ fabricator mesh reinforcement'. 29 June. Retrieved from www.youtube.com/watch?time_continue=29&v=TCJOQkOE69s

Needham, J. (1994). *The Shorter Science and Civilisation in China*. Cambridge: Cambridge University Press.

Nene, A.S. (2011). 'Rock engineering in ancient India'. Retrieved from https://gndec. ac.in/~igs/ldh/conf/2011/articles/Theme%20-%20P%202.pdf

News4 (2018). 'Blockchains L.L.C. proposes "Smart City" east of Reno'. 1 November. Retrieved from https://mynews4.com/news/local/smart-city-to-be-build-at-tahoe-reno-industrial-center

Nonaka, I. and Takeuchi, H. (1995). *The Knowledge-Creating Company: How Japanese Companies Create the Dynamics of Innovation*. Oxford: Oxford University Press.

Now Science News (2018). 'HRP-5P Humanoid Construction Robot by AIST'. 30 September. Retrieved from www.youtube.com/watch?v=qBvuZ-tUFiA

NUS News (2018). 'NUS builds new 3D printing capabilities, paving the way for construction innovations'. 5 July. Retrieved from https://news.nus.edu.sg/press-releases/construction-3D-printing

Oak Ridge National Laboratory (2017). 'Project AME'. Retrieved from https://web.ornl. gov/sci/manufacturing/projectame/

Oakland, J. and Turner, M. (2015). *Leading Quality in the 21st Century*. London: CQI. Retrieved from www.quality.org/file/494/download?token=UFcUGvXy

Panditabhushana V-Subrahmanya Sastri, B. *Brihat Samhita of Varaha Mihira* LVI.31, LVII 1–7. (Trans. 1946). Retrieved from https://archive.org/stream/Brihatsamhita/brihat-samhita_djvu.txt

Papadopoulos, K. and Vintzileou, E. (2013). 'The new "poles and empolia" for the columns of the ancient Greek temple of Apollo Epikourios'. Retrieved from www. bh2013.polimi.it/papers/bh2013_paper_229.pdf

Pliny the Younger (1900). *Letters*, Book 10, Letter 18. Trans. J.B. Firth. Retrieved from www.attalus.org/old/pliny10a.html

Plutarch (1996). *Pericles*. Trans. J. Dryden, Chapter 12. Retrieved from https://people. ucalgary.ca/~vandersp/Courses/texts/plutarch/plutperi.html#XII

PriceWaterhouseCoopers (2017). 'Sizing the prize: What's the real value of AI for your business and how can you capitalise?' Retrieved from www.pwc.com/gx/en/issues/analytics/assets/pwc-ai-analysis-sizing-the-prize-report.pdf

PriceWaterhouseCoopers (2018a). 'Will robots really steal our jobs?', p. 31, Figure 6.6. Retrieved from www.pwc.co.uk/economic-services/assets/international-impact-of-automation-feb-2018.pdf

PriceWaterhouseCoopers (2018b). 'AI will transform project management. Are you ready? Transformation assurance'. Retrieved from https://news.pwc.ch/wp-content/uploads/2018/04/AI-will-transform-PM-Whitepaper_EN_web.pdf

PriceWaterhouseCoopers 'Skies without limits'. Retrieved from www.pwc.co.uk// intelligent-digital/drones/Drones-impact-on-the-UK-economy-FINAL.pdf

ProGlove 'Ready for 4.0'. Retrieved from www.proglove.de/products/#wearables

Publius Papinius Statius (2003). *Silvae*, Book IV: 3, the Via Domitiana. Cambridge, MA: Loeb.

Quercia, G. and Brouwers, H.J.H. (2010). 'Application of nano-silica (nS) in concrete mixtures'. Retrieved from www.researchgate.net/profile/George_Quercia_Bianchi/publication/257029738_Application_of_nano-silica_nS_in_concrete_mixtures/links/00b7d5243e5e804358000000.pdf

Ransome, F. (1866). *Patent Paving Stone*. Retrieved from https://books.google.co.uk/book s?hl=en&lr=&id=66wQAQAAIAAJ&oi=fnd&pg=PA1&dq=building+quality+

inspection&ots=8_-lAY_aPy&sig=cSy6JxSr-psry0CRSIppJetT7oE#v=onepage&q=building%20quality%20inspection&f=false

Rehm, A. (1958–68). *Didyma II: Die Inschriften*. No. 48. Berlin.

Rio Tinto (2018). 'Smarter technology'. Retrieved from www.riotinto.com/ourcommitment/smarter-technology-24275.aspx

RMIT University (2016). 'How brickmakers can help butt out litter'. 27 May. Retrieved from www.rmit.edu.au/news/all-news/2016/may/how-brickmakers-can-help-butt-out-litter

Rudgley, R. (1999). *Lost Civilisations of the Stone Age*. London: Arrow Books.

Ryan, G. (2018). 'Hammond pledges £100m for National Retraining Scheme', *Times Educational Supplement*, 1 October. Retrieved from www.tes.com/news/national-retraining-scheme-philip-hammond

Sainty, J.C. (2002). *Ordnance Surveyor 1538 to 1854*. London: Institute of Historical Research. Retrieved from www.history.ac.uk/publications/office/ordnance-surveyor.

Schubmehl, D. (2014). 'Unlocking the hidden value of information'. IDC Community. 14 July. Retrieved from https://idc-community.com/groups/it_agenda/bigdataanalytics/unlocking_the_hidden_value_of_information

SkyCatch, 'All-in-one drone data solution for enterprise'. Retrieved from www.skycatch.com

Skylar Tibbits (2013). 'The emergence of "4D printing"'. TED. Retrieved from www.youtube.com/watch?time_continue=1&v=0gMCZFHv9v8

Smartvid.io 'Tap into the power of AI (artificial intelligence) to reduce risk on your projects'. Retrieved from www.smartvid.io

Smisek, P. (2017). *A Short History of 'Bricklaying Robots'*. 17 October. B1M video channel. Retrieved from www.theb1m.com/video/a-short-history-of-bricklaying-robots

Smith, C.B. (2018). *How the Great Pyramid Was Built*. London: Penguin Random House.

Snoeck, D., Van Tittelboom, K., Wang, J., Mignon, A., Feiteira, J. *et al.* (2014). 'Self-healing of concrete'. Ghent University, 27 March. Retrieved from www.ugent.be/ea/structural-engineering/en/research/magnel/research/research3/selfhealing

Société d'Exploitation de la tour Eiffel. *Origins and Construction of the Eiffel Tower*. Retrieved from www.toureiffel.paris/en/the-monument/history

Surmet's ALON®. (2011). 'Transparent Armor 50 caliber test'

Tacitus (1876). *The Annals*, Book 4, p. 62. Retrieved from https://en.wikisource.org/wiki/The_Annals_(Tacitus)/Book_4#62. Translation based on Alfred John Church and William Jackson Brodribb.

The Construction Index 'Construction pre-tax margins average 1.5%'. Retrieved from www.theconstructionindex.co.uk/news/view/construction-pre-tax-margins-average-15 (accessed 28 August 2017).

The Economist (2018). 'China has built the world's largest water-diversion project', 5 April. Retrieved from www.economist.com/china/2018/04/05/china-has-built-the-worlds-largest-water-diversion-project

The Nation (2018). 'Meeting the demands of posh condo buyers'. 9 July. Retrieved from www.nationmultimedia.com/detail/Real_Estate/30349402

Tobii Pro 'Eye tracking for research'. Retrieved from www.tobiipro.com

Trudeau, C.R. (2012). *The Public Speaks: An Empirical Study of Legal Communication*. Retrieved from https://works.bepress.com/christopher_trudeau/1/

Turing, A.M. (1950). *Computing Machinery and Intelligence*. Retrieved from https://academic.oup.com/mind/article/LIX/236/433/986238

Tybot 'Reliable, flexible, and scalable solution for bridge deck construction'. Retrieved from www.tybotllc.com

UK government (2017). 'Industrial strategy: building a Britain fit for the future'. White Paper, p. 10. CM9529. Retrieved from https://assets.publishing.service.gov.uk/government/uploads/system/uploads/attachment_data/file/730043/industrial-strategy-white-paper-print-ready-a4-version.pdf

UK Parliament (1938). 'Oral answers to questions: Housing-Building Standards'. Hansard, vol. 342, col. 594. 1 December. London: HMSO. Retrieved from https://hansard.parliament.uk/Commons/1938-12-01/debates/1a2ca5b1-e3a1-4fcb-92a7-18b9a9700d83/BuildingStandards?highlight=national%20house%20builders%27%20registration%20council#contribution-7f3d1aaf-7449-44bf-a527-5aeb28a571f1

UK Parliament (1952). 'Oral answers to questions: Housing-Building Standards', Hansard, vol. 452, col. 195. 4 March. London: HMSO. Retrieved from https://hansard.parliament.uk/Commons/1952-03-04/debates/2ebaeb17-1023-47f8-8132-82148a497f01/BuildingStandards

UK Parliament (1967). 'Oral answers to questions: Scotland. Private House Building (Standards)'. Hansard, vol. 755, col. 1415. 6 December. London: HMSO. Retrieved from https://hansard.parliament.uk/Commons/1967-12-06/debates/23b399c4-b8e2-421d-b8ee-300390901fed/PrivateHouseBuilding(Standards)

UK Parliament (1977). 'Oral answers to questions: Northern Ireland. House Building Standards', Hansard, vol. 933, col. 1729. 23 June. London: HMSO. Retrieved from https://hansard.parliament.uk/Commons/1977-06-23/debates/9eb4c8b8-63e3-49ae-a37a-54dd823f6af0/HouseBuildingStandards

Upskill, 'Augmented reality use cases in enterprise'. Retrieved from https://upskill.io/skylight/use-cases/

Van Wijnen (2018). 'World first: living in a 3D printed house made of concrete'. Retrieved from https://translate.google.com/translate?hl=en&sl=nl&u=www.vanwijnen.nl/actueel/wereldprimeur-wonen-in-een-3d-geprint-huis-van-beton/&prev=search

Vastushastraguru.com. *Vastu Shastras*. Retrieved from www.vastushastraguru.com.

Vitruvius (2001). *Ten Books of Architecture*. Cambridge: Cambridge University Press.

Volvo CE (2017a). 'LX1 prototype hybrid wheel loader delivers 50% fuel efficiency improvement'. Press release, 7 December. Retrieved from www.volvoce.com/global/en/news-and-events/news-and-press-releases/2017/lx1-prototype-hybrid-wheel-loader-delivers-50-percent-fuel-efficiency-improvement

Volvo CE (2017b). 'Volvo CE unveils the next generation of its Electric Load Carrier concept'. Retrieved from www.volvoce.com/united-states/en-us/about-us/news/2017/volvo-ce-unveils-the-next-generation-of-its-electric-load-carrier-concept/

Volvo CE 'Compact assist for asphalt with Density Direct'. Retrieved from www.volvoce.com/united-states/en-us/products/other-products/density-direct/

Volvo CE and LEGO® (2018). 'Volvo CE: Introducing ZEUX in collaboration with the LEGO® Group'. Retrieved from www.youtube.com/watch?time_continue=25&v=3uJCgt_2Y4o

Wakisaka, T., Furuya, N., Hishikawa, K., *et al.* (2000). *Automated Construction System for High-rise Reinforced Concrete Buildings*. Retrieved from www.iaarc.org/publications/fulltext/Automated_construction_system_for_high-rise_reinforced_concrete_buildings.PDF

Wanberg, J., Harper, C. and Hallowell, M.R. (2013). 'Relationship between construction safety and quality performance'. *Journal of Construction Engineering and Management* 139: 10.

Wilmore, J. (2018). 'Will Amazon enter the construction jungle?' *Construction News*. 8 May. Retrieved from www.constructionnews.co.uk/analysis/cn-briefing/will-amazon-enter-the-construction-jungle/10030732.article

Wilson, W. and Rhodes, C. (2018). 'New-build housing: construction defects: issues and solutions (England)'. London: House of Commons Library, Retrieved from research-briefings.files.parliament.uk/documents/CBP-7665/CBP-7665.pdf

Witzel, M. and Warner, M. (eds) (2013). *The Oxford Handbook of Management Theorists.* Oxford: Oxford University Press.

World Economic Forum, (2018). *Future Scenarios and Implications for the Industry.* Retrieved from www3.weforum.org/docs/Future_Scenarios_Implications_Industry_report_2018.pdf

Xiyi, L. (c. 1235) *Kao gong ji.* Trans. 2013. New York: Routledge.

XOi Technologies 'It starts in the field'. Retrieved from www.xoi.io/solution/

Yhnova (2017). 'A robot 3D printer is building a house in Nantes'. Retrieved from http://batiprint3d.fr/en/

Zhu, H. (2015). 'BIM: A "model" method'. The BIM hub. 17 October. Retrieved from https://thebimhub.com/2015/10/17/bim-a-model-method/#.W8R6Oy_MxQI

Further reading

Atkinson, G. (1987). *A Guide through Construction Quality Standards*. Wokingham: Van Nostrand Reinhold.

Barnes, H. (2006). *Gannibal: The Moor of Peterburg*. London: Profile Books Ltd.

Barrat, J. (2013). *Our Final Invention: Artificial Intelligence and the End of the Human Era*. New York: Thomas Dunne Books.

Broadbent, M. and Kitzis, E.S. (2005). *The New CIO Leader*. Boston: Harvard Business School Press.

Brynjolfsson, E. and McAfee, A. (2014). *The Second Machine Age*. New York: W.W. Norton & Co.

Child, M. (2000). *Discovering Church Architecture: A Glossary of Terms*. Princes Risborough: Shire Publications Ltd.

Chudley, R. and Greeno, R. (2008). *Building Construction Handbook*, 7th edn. Oxford: Butterworth-Heinemann.

Crosby, P.B. (1979). *Quality Is Free*. Maidenhead: McGraw-Hill.

Dekker, S. (2006). *The Field Guide to Understanding Human Error*. Farnham: Ashgate.

Feigenbaum, A.V. (1991). *Total Quality Control*, 3rd edn. New York: McGraw-Hill.

Ferguson, I. and Mitchell, E. (1986). *Quality on Site*. London: Batsford.

Finlay, S. (2017). *Artificial Intelligence and Machine Learning for Business*. London: Relativistic Books.

Ford, M. (2016). *The Rise of the Robots*. London: Oneworld Publications.

Goetsch, D.L. and Davis, S.B. (1994). *Quality Management*, 5th edn. Upper Saddle River, NJ: Pearson.

Gomes, H. (1996). *Quality Quotes*. Milwaukee, WI: ASQC Quality Press.

Goodwin, T. (2018). *Digital Darwinism*. London: Kogan Page.

Harari, T.N. (2011). *Sapiens: A Brief History of Mankind*. London: Penguin Random House LLC.

Harari, T.N. (2015). *Homo Deus: A Brief History of Tomorrow*. London: Penguin Random House LLC.

Harris, F. and McCaffer, R. (2013). *Modern Construction Management*, 7th edn. Chichester: Wiley-Blackwell.

Harris, R. (2001). *Discovering Timber-Framed Buildings*. Princes Risborough: Shire Publications Ltd.

Hill, D. (1984). *A History of Engineering in Classical and Medieval Times*. New York: Barnes & Noble Inc.

Imai, M. (1986). *Kaizen: The Key to Japan's Competitive Success*. New York: McGraw-Hill.

Jackson, N. and Dhir, R.K. (1996). *Civil Engineering Materials*, 5th edn. New York. Palgrave.

Juran Foundation Inc. (1995). *A History of Managing for Quality*. Milwaukee, WI: ASQC Quality Press.

Juran Institute Inc. (2016). *Juran's Quality Handbook: The Complete Guide to Performance Excellence*, 7th edn. London: McGraw-Hill Education.

Kelly, K. (2016). *The Inevitable: Understanding the 12 Technological Forces That Will Shape Our Future*. New York: Penguin Random House LLC.

Landels, J.G. (1978). *Engineering in the Ancient World*. London: Constable.

Nonaka, I. and Takeuchi, H. (1995). *The Knowledge-Creating Company: How Japanese Companies Create the Dynamics of Innovation*. Oxford. Oxford University Press.

O'Brien, J.J. (1974). *Construction Inspection Handbook*. New York: Van Nostrand Reinhold Co.

Oakland, J. and Marosszeky, M. (2017). *Total Construction Management: Lean Quality in Construction Project Delivery*. London: Routledge.

Ross, A. (2016). *The Industries of the Future*. London: Simon & Schuster.

Rudgley, R. (1999). *Lost Civilisations of the Stone Age*. London: Arrow Books.

Susskind, R. and Susskind, D. (2015). *The Future of the Professions*. Oxford: Oxford University Press.

Tegmark, M. (2017). *Life 3.0*. New Delhi: Penguin Random House LLC.

Tierney, T.F. (2016). *Intelligent Infrastructure*. Charlottesville, VA: University of Virginia Press.

West, T.W. (2000). *Discovering English Architecture*. Princes Risborough: Shire Publications Ltd.

Wright, G.N. (2004). *Discovering Abbeys and Priories*: Princes Risborough. Shire Publications Ltd.

Standards

You should be aware that British and International standards are reviewed and republished frequently. Check BSI and ISO websites for the latest versions.

BSI BS 1192:2007 + A2:2016 *Collaborative Production of Architectural, Engineering and Construction Information: Code of Practice*. Milton Keynes: BSI Standards Limited.

BSI BS 1192–4:2014 *Collaborative Production of Information: Fulfilling Employer's Information Exchange Requirements Using COBie. Code of Practice*. Milton Keynes: BSI Standards Limited.

BSI BS EN ISO 9000:2015 *Quality Management Systems: Fundamentals and Vocabulary*. Milton Keynes: BSI Standards Limited.

BSI BS EN ISO 9001:2015 *Quality Management Systems: Requirements*. Milton Keynes: BSI Standards Limited.

BSI BS EN ISO 14001:2015 *Environmental Management Systems: Requirements with Guidance for Use*. Milton Keynes: BSI Standards Limited.

BSI BS EN ISO 19650–1 *Organization of Information About Construction Works – Information Management Using Building Information Modelling: Part 1: Concepts and principles*. (due for publication in 2018). Milton Keynes: BSI Standards Limited.

BSI BS EN ISO 19650–2 *Organization of Information About Construction Works – Information Management Using Building Information Modelling: Part 2: Delivery Phase of Assets*. Milton Keynes: BSI Standards Limited.

BSI BS EN ISO/IEC 27001:2017 *Information Technology. Security Techniques. Information Security Management Systems. Requirements.* Milton Keynes: BSI Standards Limited.

BSI BS ISO 10005:2018 *Quality Management: Guidelines for Quality Plans.* Milton Keynes: BSI Standards Limited.

BSI BS ISO 10006:2017 *Quality Management: Guidelines for Quality Management in Projects.* Milton Keynes: BSI Standards Limited.

BSI BS ISO 10015:1999 *Quality Management: Guidelines for Training.* Milton Keynes: BSI Standards Limited.

BSI BS ISO 10018:2012 *Quality Management: Guidelines on People Involvement and Competence.* Milton Keynes: BSI Standards Limited.

BSI BS ISO 30301–2011 *Information and Documentation: Management Systems for Records – Requirements.* Milton Keynes: BSI Standards Limited.

BSI BS ISO 30302–2015 *Information and Documentation: Management Systems for Records – Guidelines for Implementation.* Milton Keynes: BSI Standards Limited.

BSI BS ISO 31000–2018 *Risk Management: Guidelines.* Milton Keynes: BSI Standards Limited.

BSI BS ISO 45001:2018 *Occupational Health and Safety Management Systems: Requirements with Guidance for Use.* Milton Keynes: BSI Standards Limited.

BSI PAS 1192–2:2013 *Specification for Information Management for the Capital/Delivery Phase of Construction Projects Using Building Information Modelling.* Milton Keynes: BSI Standards Limited.

BSI PAS 1192–3:2014 *Specification for Information Management for the Operational Phase of Assets Using Building Information Modelling (BIM).* Milton Keynes: BSI Standards Limited.

BSI PAS 1192–5:2015 *Specification for Security-Minded Building Information Modelling: Digital Built Environments and Smart Asset Management.* Milton Keynes: BSI Standards Limited.

BSI PAS 1192–6:2018 *Specification for Collaborative Sharing and Use of Structured Health and Safety Information Using BIM.* Milton Keynes: BSI Standards Limited.

BSI PD 7504–2005 *Knowledge Management: Public Sector: A Guide to Good Practice.* Milton Keynes: BSI Standards Limited.

BSI PD 7505–2005 *Skills for Knowledge Working: A Guide to Good Practice.* Milton Keynes: BSI Standards Limited.

BSI PD 7506–2005 *Linking Knowledge Management with Other Organizational Functions and Disciplines: A Guide to Good Practice.* Milton Keynes: BSI Standards Limited.

Index